Deepen Your Mind

Deepen Your Mind

# 推薦序

　　大家好，我是在中部舉辦聊天機器人小小聚的佳新，本身在彰化開軟體公司，提供數位轉型解決方案。作者 Wolke 跟我除了都是 LINE 官方認證的技術專家（LINE API Expert，簡稱 LAE）之外，我們也都是彰化鄉親。他曾經支援過小聚好幾次開講，大老遠跑來台中的聚會場地，帶領工作坊。

　　我發現 Wolke 的實務能力很強，教學技巧也很純熟，而且非常樂於分享和參加社群活動，不愧是頂著雙料技術專家頭銜的大神級人物。他除了是 LAE 之外，也是 Google 官方認證的技術專家（Google Developer Expert，簡稱 GDE）。

　　果然後來也發現他在線上開設了幾門課，而如今，還誕生了一本電腦書，也就是大家手頭上的這本《LINE 聊天機器人 +AI+ 雲端 + 開源＋程式：輕鬆入門到完整學習》。我有幸在第一時間拜讀，不得不說，在這個可能是有史以來最強調程式設計和運算思維的 108 課綱年代，JavaScript 仍不失為非常好的初學者程式語言。

　　因為工作的緣故，我經常被要求推薦程式語言和電腦書，現在我就要告訴各位家長、老師、同學、讀者，Wolke 這本書之所以也適合你的 5 個原因。

一、深入淺出，從基本語法、開發工具開始教起，無痛入門。

二、適合當作課本，章節的最後會有幾題有趣又有深度的練習。

三、完整範例，跟著做就可以完成一個專案（猜拳遊戲夠簡單了吧）。

四、領先時代，LINE 聊天機器人、web3/NFT 區塊鏈也都有涉獵喔。

五、大家都在用公有雲，東西都上雲，有教怎麼使用 Google Cloud。

　　期待閱讀完這篇推薦序之後，你會毫不猶豫地拿著這本書走向結帳櫃臺，接著在短則數週、長則數個月的時間之後，在浩瀚的網際網路世界的某個地方，可以看到你應用。到時候，也歡迎分享給我！

陳佳新 于 2022 年秋天
jarsing@chibuapp.com
奇步應用有限公司 執行長
LINE API Expert & Microsoft AI MVP

# 目錄

## ▌0　本書使用指南

0-1　感謝你打開這本書 .................................................................0-2

　　0-1-1　　自我介紹 .................................................................0-2

0-2　本書大致可以著重在三個面向的學習重點 .............................0-4

　　0-2-1　　程式語法的學習 .......................................................0-4

　　0-2-2　　程式應用的開發 .......................................................0-5

　　0-2-3　　程式系統的維運 .......................................................0-5

0-3　為什麼之前學程式會失敗 ....................................................0-5

　　0-3-1　　身為初學者，一開始選擇了太難的程式語言 ...............0-5

　　0-3-2　　不好的程式學習體驗 ................................................0-6

　　0-3-3　　時間 ......................................................................0-7

0-4　為什麼以本書學程式會成功 .................................................0-7

　　0-4-1　　較友善的程式語言 ...................................................0-7

　　0-4-2　　較好的學習體驗 .......................................................0-8

　　0-4-3　　不需要擔心時間 .......................................................0-8

## ▌1　JS 程式學習篇：基本語法練習

1-1　開發環境建置 ....................................................................1-2

1-2　學習目標／演練成果 ...........................................................1-4

1-3　程式碼是什麼？ ..................................................................1-4

1-4　敘述句 Statement ...............................................................1-4

1-5　註解 ..................................................................................1-5

　　1-5-1　　註解練習 ................................................................1-6

1-6　關鍵字 / 保留字 ..................................................................1-7

1-7    變數 ..................................................................................1-7

    1-7-1    變數宣告練習 ......................................................1-8

1-8    資料型別 Data Type ........................................................1-10

    1-8-1    字串的宣告方式： ...........................................1-11

    1-8-2    資料形別練習 ....................................................1-11

1-9    變數命名規則 ................................................................1-12

    1-9-1    駝峰命名法 Camel-Case ..................................1-13

    1-9-2    變數命名規則練習 ...........................................1-13

1-10    陣列 Array ......................................................................1-14

    1-10-1    陣列練習 ..........................................................1-15

1-11    運算式＆運算子 ...........................................................1-16

    1-11-1    運算式＆運算子練習 ....................................1-18

## ▌2　IDE 篇

2-1    學習目標／演練成果 .......................................................2-2

2-2    IDE 是什麼？ ....................................................................2-2

2-3    codesandbox 可以做什麼？不能做什麼？ ....................2-2

    2-3-1    codesandbox 的好處是什麼： ...........................2-2

    2-3-2    codesandbox 的缺點： .......................................2-3

2-4    codesandbox 設定 .............................................................2-4

2-5    package.json 簡介 .............................................................2-8

    2-5-1    安裝第一個套件 nodemon ................................2-9

    2-5-2    變更版本號 ......................................................2-14

    2-5-3    移除套件 nodemon .........................................2-15

2-6    執行 nodemon ..................................................................2-15

2-7    codesandbox 的狀況排除 ..............................................2-18

    2-7-1    無法正常運行 ..................................................2-18

    2-7-2    IDE 練習 ...........................................................2-19

# 3 程式學習篇：函式方法與物件

3-1　開發環境 .................................................................. 3-2

3-2　學習目標／演練成果 ................................................ 3-2

3-3　函式 & 箭頭函式 ...................................................... 3-2

　　3-3-1　宣告及呼叫 ................................................... 3-2

　　3-3-2　解釋 node http server 的函數 ...................... 3-5

　　3-3-3　函式練習 ....................................................... 3-6

3-4　變數作用範圍／變數作用域 scope ........................ 3-9

　　3-4-1　let 宣告 ......................................................... 3-9

　　3-4-2　函式的變數作用範圍練習 ......................... 3-14

3-5　物件 ........................................................................ 3-14

　　3-5-1　宣告方式 ..................................................... 3-15

　　3-5-2　物件在實務上的使用為： ......................... 3-15

　　3-5-3　呼叫方式 ..................................................... 3-16

3-6　物件與類別 ............................................................ 3-17

　　3-6-1　類別的宣告方式 ......................................... 3-17

　　3-6-2　撰寫溫度換算 API ....................................... 3-18

　　3-6-3　物件練習 ..................................................... 3-20

3-7　全域物件 ................................................................ 3-21

　　3-7-1　關於 this ....................................................... 3-22

3-8　全域物件 STRING 字串 ......................................... 3-23

　　3-8-1　宣告 String 物件 .......................................... 3-23

　　3-8-2　全域物件 STRING 練習 ................................ 3-24

3-9　全域物件 NUMBER .................................................. 3-27

　　3-9-1　宣告為 Number 物件 ................................... 3-27

3-10　常用屬性及內建函式 ........................................... 3-27

3-11　全域物件 MATH .................................................... 3-27

3-12　常用屬性及內建函式 ........................................... 3-27

　　　3-12-1　撰寫樂透 API ..................................................3-28

　　　3-12-2　全域物件 Math 及 Number 的練習 ..................3-29

3-13　全域物件 DATE & TIME ...............................................3-29

　　　3-13-1　DATE 宣告方式 .................................................3-29

　　　3-13-2　Date 練習 ..........................................................3-34

3-14　陣列函式 ...........................................................................3-36

　　　3-14-1　陣列宣告方式 .....................................................3-37

　　　3-14-2　陣列練習 .............................................................3-38

3-15　自訂全域物件 & module ...............................................3-38

　　　3-15-1　exports 練習 ......................................................3-40

# ▎4　JS 程式學習篇：決策與迴圈

4-1　學習目標／演練成果 ........................................................4-2

4-2　條件判斷式 .........................................................................4-2

　　　4-2-1　評估條件 .............................................................4-2

　　　4-2-2　if 條件判斷式 .......................................................4-4

　　　4-2-3　switch 條件判斷式 ...............................................4-6

　　　4-2-4　捷徑／簡寫 ..........................................................4-7

　　　4-2-5　真值與假值 ..........................................................4-8

　　　4-2-6　捷徑及真假值練習 ..............................................4-9

4-3　迴圈 .....................................................................................4-9

　　　4-3-1　for 迴圈 ...............................................................4-10

　　　4-3-2　while 迴圈 ...........................................................4-11

　　　4-3-3　do while 迴圈 ......................................................4-12

　　　4-3-4　迴圈練習 ..............................................................4-13

4-4　陣列函式的迴圈應用 ........................................................4-13

　　　4-4-1　map ......................................................................4-14

　　　4-4-2　filter .....................................................................4-15

4-4-3　　reduce ....................................................................4-16

4-4-4　　sort 排序 ..............................................................4-16

4-4-5　　陣列函式的迴圈應用練習 ..........................................4-18

# ▌5　JS 程式學習篇：非同步與 API

5-1　學習目標／演練成果 ........................................................5-2

5-2　同步 sync vs 非同步 async ................................................5-2

　　5-2-1　　何謂同步？ ...........................................................5-2

　　5-2-2　　何謂非同步？ ........................................................5-3

　　5-2-3　　阻塞（blocking） ..................................................5-3

　　5-2-4　　單線程 single threaded ...........................................5-4

　　5-2-5　　堆疊追蹤 ..............................................................5-5

　　5-2-6　　非同步語法 (setTimeout) .......................................5-6

　　5-2-7　　AJAX( Asynchronous JavaScript and XML) .................5-8

5-3　非同步語法的實現 ...........................................................5-8

　　5-3-1　　XMLHttpRequest ...................................................5-9

　　5-3-2　　Fetch ...................................................................5-9

　　5-3-3　　Async & Await ....................................................5-10

5-4　Promise 物件與 async & await 語法 ................................5-10

　　5-4-1　　Promise 物件宣告方式 ..........................................5-10

　　5-4-2　　await & async 宣告方式 .........................................5-13

　　5-4-3　　Async & Await & Promise 練習 ...............................5-14

5-5　並行運算 concurrent computing vs 平行運算 parallel computing ...5-14

# ▌6　實作練習篇：實作網站版剪刀石頭布

6-1　學習目標／演練成果 ........................................................6-2

6-2　使用者輸入參數 ..............................................................6-2

6-3　新增 GAME 物件 .............................................................6-4

6-4　撰寫 GAME 物件 ...................................................... 6-6

　　6-4-1　宣告常數 .......................................................... 6-6

　　6-4-2　撰寫遊戲邏輯 .................................................. 6-7

　　6-4-3　紀錄遊戲的輸贏：探討 閉包 Clourse ............ 6-13

　　6-4-4　class vs clourse 想一想 ................................. 6-21

6-5　遊戲邏輯程式碼放在前端 ...................................... 6-22

　　6-5-1　觀察一下，程式放在前後端的差別 ............. 6-26

　　6-5-2　避免巢狀結構 .................................................. 6-26

6-6　使用 framework 框架 express ................................ 6-27

　　6-6-1　Hello World ...................................................... 6-29

　　6-6-2　路由 route ........................................................ 6-32

　　6-6-3　使用 express 改寫 剪刀 石頭 布 ................... 6-32

　　6-6-4　練習 ................................................................... 6-36

　　附帶一提：使用 webpack 打包 ................................. 6-37

　　說明 ............................................................................. 6-39

## 7　開源篇：npm & github

7-1　學習目標／演練成果 ................................................ 7-2

7-2　實做開源套件：天氣小幫手 ................................... 7-2

　　7-2-1　申請中央氣象局資料開放平台 ..................... 7-2

　　7-2-2　撰寫 ................................................................... 7-8

7-3　改寫成可以發佈的形式 ........................................... 7-11

　　7-3-1　發佈 NPM 套件 ................................................ 7-14

　　7-3-2　測試剛剛發佈的套件 ..................................... 7-18

7-4　GitHub 發佈 .............................................................. 7-19

　　7-4-1　註冊 github ....................................................... 7-19

　　7-4-2　上傳 github ....................................................... 7-20

7-5　加一個 nodeman 避免 codesandbox 錯認為 前端開發環境 ............7-23

7-6　Link Sandbox ............................................................ 7-24

7-7  加上 github repository 跟 npm 上的 package 的關聯性 ...................7-27
    7-7-1  改寫 package.json .......................................................7-28
    7-7-2  NPM 再發佈一版 ........................................................7-29
    7-7-3  改寫 README.md .......................................................7-30
    7-7-4  更新 github .............................................................7-32
7-8  協作開發 ..........................................................................7-33
    7-8-1  透過 github 開 sandbox ...............................................7-33
    7-8-2  發 PR pull request .....................................................7-35
    7-8-3  合併 PR pull request ..................................................7-38
    7-8-4  小結 .......................................................................7-39
7-9  Open Source 開源生態圈 ...................................................7-39
    7-9-1  開源的好處 .............................................................7-40
    7-9-2  開源的挑戰 .............................................................7-40
    7-9-3  小結 .......................................................................7-43
    7-9-4  開源相關的工具及平台 .............................................7-44
    7-9-5  開源練習 .................................................................7-44

# 8 LINE Bot 篇：LINE Message API

8-1  前言 ................................................................................8-2
8-2  學習目標／演練成果 .........................................................8-2
8-3  實做 天氣小幫手 LINE bot 版 .............................................8-2
    8-3-1  申請 LINE Bot 開發者帳號 .........................................8-2
    8-3-2  建置 LINE Bot 伺服器 ...............................................8-9
    8-3-3  解密 .......................................................................8-12
    8-3-4  回傳訊息 .................................................................8-16
    8-3-5  引用 天氣小幫手套件 ...............................................8-20
8-4  小結 ................................................................................8-25
8-5  常用功能解釋 ...................................................................8-27
    8-5-1  webhook ..................................................................8-27

| | | |
|---|---|---|
| 8-5-2 | client | 8-28 |
| 8-5-3 | 中介軟體 middleware | 8-29 |
| 8-5-4 | push & reply api | 8-29 |
| 8-5-5 | Events | 8-30 |
| 8-5-6 | Message objects | 8-32 |
| 8-5-7 | Rich menu | 8-34 |
| 8-5-8 | Quick Reply | 8-37 |
| 8-5-9 | Flex Message | 8-38 |
| 8-5-10 | Actions | 8-41 |
| 8-5-11 | Groups Chat | 8-42 |
| 8-6 | 附帶一提：程式工程師開發新內容的起手式 | 8-43 |
| 8-6-1 | 尋找 LINE Bot 的開發資源 | 8-43 |

# ▊ 9　機器學習篇：Dialogflow

| | | |
|---|---|---|
| 9-1 | 前言 | 9-2 |
| 9-2 | 學習目標／演練成果 | 9-2 |
| 9-3 | dialogflow 基本介面操作 | 9-3 |
| 9-3-1 | 使用 dialogflow | 9-3 |
| 9-3-2 | dialogflow 串接 LINE Bot | 9-7 |
| 9-4 | Intents 意圖 | 9-11 |
| 9-5 | Entities | 9-14 |
| 9-6 | call dialogflow API | 9-17 |
| 9-6-1 | codesandbox 環境建置 | 9-17 |
| 9-6-2 | 拿憑證 | 9-18 |
| 9-6-3 | 使用 dialogflow API | 9-26 |
| 9-7 | dialogflow 練習一 | 9-29 |
| 9-7-1 | 改成 npm 套件並發佈 | 9-29 |
| 9-8 | dialogflow 練習二 | 9-33 |
| 9-8-1 | 結合 LINE Bot 完成天氣小幫手 | 9-33 |

9-9　　dialogflow 練習三 ......................................................9-38

　　　9-9-1　　利用 dialogflow API 將 剪刀石頭布的遊戲邏輯 加進天氣
　　　　　　　小幫手 LINE Bot 裡，又不影響原有的天氣查詢功能...9-38

# ▌ 10　上雲篇：GCP

10-1　前言 ............................................................................10-2

10-2　學習目標／演練成果 ....................................................10-2

10-3　部屬到 Google Cloud Function ....................................10-2

　　　10-3-1　　啟用 google cloud function ........................10-2

　　　10-3-2　　將 LINE Bot 放上 google cloud function ......10-7

　　　10-3-3　　上線天氣小幫手 ....................................10-12

10-4　附帶一提：後端解決方案歷史 ....................................10-17

　　　10-4-1　　本機機房 ............................................10-17

　　　10-4-2　　代管機房 ............................................10-17

　　　10-4-3　　PAAS (platform as a service) 平台即服務 ....10-18

　　　10-4-4　　Microservices 微服務 ............................10-18

　　　10-4-5　　FAAS Function-as-a-Service 功能即服務 ......10-18

10-5　練習 ..........................................................................10-19

# ▌ 11　總結篇：JS 的無限可能

11-1　後端解決方案 ............................................................11-2

　　　11-1-1　　Website Backend ..................................11-2

　　　11-1-2　　Chatbot Backend ..................................11-3

　　　11-1-3　　Web API ..............................................11-4

　　　11-1-4　　GraphQL ............................................11-4

11-2　前端網頁 ....................................................................11-5

　　　11-2-1　　React ..................................................11-5

　　　11-2-2　　Vue....................................................11-6

　　　11-2-3　　angular ..............................................11-7

11-3　區塊鏈 .................................................................................... 11-7

    11-3-1　web3.js ....................................................................... 11-7

    11-3-2　NFT ........................................................................... 11-8

11-4　人工智慧 ................................................................................ 11-8

    11-4-1　tensorflow.js .............................................................. 11-9

    11-4-2　人工智慧相關服務 API .............................................. 11-9

11-5　IOT ......................................................................................... 11-9

    11-5-1　Node-RED ................................................................. 11-9

11-6　APP 行動應用程式 .............................................................. 11-10

    11-6-1　React Native ............................................................ 11-10

11-7　桌面應用軟體 ...................................................................... 11-11

    11-7-1　Electron .................................................................... 11-11

11-8　自動化測試及爬蟲 .............................................................. 11-12

    11-8-1　selenium-webdriver .................................................. 11-13

    11-8-2　puppeteer ................................................................. 11-13

11-9　其他 ...................................................................................... 11-14

11-10 以 JS 為基礎發展的語法等 ................................................ 11-14

    11-10-1　Google Apps Script ................................................ 11-14

    11-10-2　Google Ads scripts ................................................ 11-14

    11-10-3　TypeScript ............................................................. 11-15

    11-10-4　JSON ...................................................................... 11-15

# 0

## 本書使用指南

## ▌ 0-1　感謝你打開這本書

感謝你打開這本書，我不知道你的動機為何？

我猜想，可能你的身分或是你的工作，迫使你必須學習一些，又或是很多程式上的技巧，來滿足你的專業或是學業上的需求，所以你打開了這本書。

又或者，你只是因為單純的：「興趣」，又剛好有時間，在這個閒暇的午後裡，書架上，發現我這一本有趣的書。

簡簡單單就能學會這所有的程式技能？

這有可能嗎？

畢竟在那之前，你也曾是花過時間，費過苦功，練習過一番，但很明顯，成效不彰，所以你現在翻起了這本書。

而我對打開這本書的你，頗為好奇，如果你願意的話，可以讓我邀請你到我的 medium 上留言，我想知道確切的原因，使你找到本書之後，打開了本書。

https://wolkesau.medium.com/

忘了自我介紹一番。

### 0-1-1　自我介紹

我是一位程式設計師， programer , coder ，又或者，我更希望介紹自己為一位 developer 開發者，因為我們程式圈裡的總喜歡這麼稱呼。

你除了可以在我的 medium 上，發現我的文章之外，也會發現我活躍在程式社群，又或是開源專案，因為喜歡分享，所以我在 2018 年成為了 LAE ，2019 年成為了 GDE。

## \* LAE: LINE Message API Expert

左下第 3 位那位瘦子

## \* GDE: Google Developer Expert

右下第 4 位那位胖子

這是全球程式圈 hero program ，能成為 LAE 及 GDE 是得經過重重關卡的，人數也是少之又少，而我有幸成為台灣代表之一。

也因為這個緣故，我在 2019 年時，在全台辦了 30 場的程式工作坊，走訪了許多地方，例如六都的都會區，或者是較為二線的區域，諸如：新營，花蓮，屏東等等；到過這些地方，和許多同好牽起了緣份。

舉辦的程式工作坊以入門為主，除了我意料之中的大學生、社會人士之外，我意料之外、意想不到的參與者也會出現，例如剛忙完農事的農婦，國小 4 年級的同學，耄耋之年的退休長者等。

和上千開發者，實體面對面的交流，使我收益良多，也更理解學習程式專業上，會有哪些問題與盲點，當年便在心中醞釀著本書的誕生。

原本 2020 年還要繼續這個路線辦理，目標是是一百場的工作坊，可惜因疫情來攪局的關係，只能轉為線上；但線上的活動，就是少了…「溫度」，沒了「溫度」，也就少了「動力」。

終於，在深智出版社的邀請之下，重新燃起「動力」，撰寫本書。

本書是總結筆者觀察到很多人在學習程式上的問題，所寫出來的一本書，本書的目地是，盡量的讓初學者在專業能力不足的情況，能夠依照著本書的教學，一窺程式業界的面目。

## 0-2 本書大致可以著重在三個面向的學習重點

1. 程式語法的學習
2. 程式應用的開發
3. 程式系統的維運

### 0-2-1 程式語法的學習

基礎的程式語法是很重要的，本書前面篇幅偏重在語法的學習與練習。

例如：JS 程式學習篇：基本語法練習、JS 程式學習篇：函式方法與物件、JS 程式學習篇：決策與迴圈、JS 程式學習篇：非同步與 API

## 0-2-2 程式應用的開發

懂了程式語法之後，就要有辦法使用程式語法以及使用幫助開發的工具，建構出符合需求的應用程式。

本書在：IDE 篇、實作練習篇：實作網站版剪刀石頭布、開源篇：npm & github、LINE Bot 篇：LINE Message API，都有實作及探討。

## 0-2-3 程式系統的維運

程式要如何上線，上線之後如何持續維護，在：開源篇：npm & github、上雲篇：GCP，有實作及探討。

# 0-3 為什麼之前學程式會失敗

本書會這樣規畫，也是來自於北中南的教學經驗中，我所看到初學者問題，總括有三：

1.  身為初學者，一開始選擇了太難的程式語言
2.  不好的程式學習體驗
3.  時間

## 0-3-1 身為初學者，一開始選擇了太難的程式語言

在辦工作坊時，最常聽到的就是 JS 真的比他們以前在大學裡學的語言，好學許多。

讓我想起筆者以前學程式時，因為沒有太多程式語言可以選，懂得程式語言的老師也不多，而他們熟悉的語言，也就那幾種。

筆者非本科系出身，出社會之後，因為總覺得這不是自己喜歡的工作，故本著對程式的興趣，毅然而然的離職進修去了。

尤記得筆者當年報名的是某知名機構，在只有半年的程式專班，該程式班教授的程式語言是 C++ ，在毫無基礎的情況之下學習 C++ ，那真的是相當的辛苦與痛苦，在熬過了那半年，到後來重新進入職場，開始撰寫程式後，學習了 VB, JAVA , JavaScript ，理解到了 C 系列語言，真的是比較難。

## 0-3-2　不好的程式學習體驗

除此之外，該程式專班，在只有半年裡，還塞入了的大學四年的程式專業課程，尤記得當時，我們的課程是以一周六天，早中晚一直上，完全沒有任何的緩衝時間，現在回頭來看，這真的是太惡搞了。

你可以想見，當時的筆者，每天的學習就是活在挫折痛苦之中，尤其是一想到繳了快 20 萬的學費，那可是我當時快半年的薪水，我是帶著破釜沉舟的心情，辭掉工作去報名的，在那樣的情況之下，也只能硬著頭皮學下去。

因為這樣的課程安排，所以陣亡率非常之高，同期進來者，待到最後連我剩 3 個，而另外 2 個待過業界寫過程式，是回來的進修的，所以那其實是算是進修班，可能是因為湊不滿學生，就這麼惡搞加了一些基礎課程就來招初學者了。

筆者不是相關科系畢業，當年的大學也沒有像現在這麼重視 IT 相關的知識，身為一個入門者，在當下真的看不出來，在只有半年時間可以學習的情況下，這樣的課程安排是有問題的，就這樣一知半解的把頭給洗下去了，真是非常的憨膽。

我還記得那時同班還有位業界企畫 ，因為不爽程式工程師常把他的企畫打槍，所以申請了在職進修進來學，然後兩週後，他就決定老老實實的回去做他的企畫了。

現在回過來頭來看，其實不是程式語言太難學，而是整個學習規畫有問題。

### 0-3-3　時間

程式工程師在撰寫程式時，是處於相當專心的狀態，也會很享受這樣的狀態之下，且工作效率也非常高，一寫 8 個小時，也不是問題，而現在有個專有名詞，叫做心流狀態 flow。

任何的活動，不管是運動學習，都可以有心流狀態；而在心流狀態之下，最害怕就是打擾，進而打斷。

而寫程式要進入心流狀態，比其他的活動困難許多；例如，畫畫，只需要拿到畫筆就能再次進入心流狀態；但是寫程式的前置作業太多，例如，開機，設定好開發環境，讀一下之前的工作紀錄，為了接下來的開發，要讀一下 SDK，再來看一看之前寫的 code 等等，所以寫程式進入心流狀態是一件較為麻煩的事。

那寫程式進入心流狀態已經這麼麻煩了，程式的學習當然更是，如果你參加了坊間的補習班，但因為時間的關係，有一堂沒一堂的參與，能夠吸收多少，我是抱持著懷疑的態度。

## ▌0-4　為什麼以本書學程式會成功

綜合上述三個原因，在本書呈現的因應對策為：

1.　較友善的程式語言
2.　較好的學習體驗
3.　不需要擔心時間

### 0-4-1　較友善的程式語言

JS 本身在當年就是因為美術設計師有開發網頁的需求，由程式人員設計給美術人員學習使用的，所以 JS 比較好學是無庸置疑的。

## 0-4-2 較好的學習體驗

小朋友學東西，都是邊做邊玩邊學的，做好之後，就分享給爸媽來藉此獲得成就感。

故筆者參考這份教養上所獲得的經驗，本書有相當多的實作，可以讓讀者寫出來之後，分享給朋友，達到邊做邊玩邊學邊分享並獲得成就感的愉悅學習體驗。

## 0-4-3 不需要擔心時間

因為是一本書，所以你不需要擔心時間上無法配合。

想學的時候，打開；累了，就闔上，休息。

如果還是不夠的話，提示一下，筆者另有和知名線上教育平台有合作錄製線上課程，歡迎參考。

https://hiskio.com/courses/343/about

https://hiskio.com/courses/345/about

# 1

## JS 程式學習篇
## 基本語法練習

## ▌ 1-1 開發環境建置

JS 的 基本語法，僅需要使用 chrome 瀏覽器內建的主控台 console 去操作練習即可。

打開內建 console 的步驟：

1.  打開 chrome

2.  按 F12

3.  選 console 主控台

4.  > 後就可以作 JS 程式的練習

或是

1.  打開 chrome

2.  點

3.  更多工具

4. 開發人員工具

5. 選 console 主控台

6. > 後就可以作 JS 程式的練習

## ▌ 1-2 學習目標／演練成果

從這裡開始，建議可以將所有程式碼的部份，都寫打進 console 的編輯區內，有助於增加程式撰寫的熟悉度。

## ▌ 1-3 程式碼是什麼？

程式碼就是給電腦理解你希望它怎麼做並達成目的過程或是稱作路徑。

例如：食譜給了廚師，就能燒出一桌好菜；食譜裡會寫需要準備的材料、設備，煮食的過程等。

所以程式設計師寫程式碼給電腦，某方面來說，就像煮菜一般，只是這次掌廚的，不是媽媽，而是電腦。

## ▌ 1-4 敘述句 Statement

程式最小的獨立單元就是敘述句。

但這是教科書的寫法，一般人描述，通常會說第幾行程式，較少聽說有人說第幾行敘述句。

而一支程式，就是由好幾行的敘述句所構成，而每一行敘述句，在 JS 的表達方式中，會以分號；作為結束符號，但是 JS 其實可以不加分號就結束一行敘述句，但強烈建議還是加；比較好。

```
let a = 1
```

```
let a = 1;
```

在 JS 裡，並沒有嚴謹的結束規定，所以上面兩個表達式都可以。

但 codesandbox 在存檔時，會貼心的幫你加上；，藉以增加程式的可讀性。

## 1-5 註解

註解顧名思義就是，註釋、解釋的縮寫，本身的存在，不做任何的程式動作，而是因為說明程式碼的用途，而存在於程式檔案裡。

註解在今時今日的用途非常廣範，也不僅僅侷限於說明的用途；通常在程式撰寫時，為了找出 bug ，常常會將某段程式碼先註解掉，讓程式運行先乾淨點，藉以找出藏在裡面的 bug。

在 JS 的表現方式有兩種：

```
// 說明 ...
```

```
/*
說明 ...
*/
```

兩種方式的差別在於 // 只能一行一行註解，而 /* */ 則可以包覆多行。

當註解使用在說明程式碼的用途上時，其撰寫的方式，有的人很嚴謹，有的人很隨性，但筆者以為重點只有一個：要讓你自己跟接手的人能看的懂。

例如：

1.  你的英文沒有很好，就不用硬寫英文，導致日後自己與同事的麻煩。
2.  盡量不要在註解區裡宣洩你的情緒，誰知道日後老闆會不會去看 code 呢。
3.  JS 的習慣，通常 release 版，會將註解拿掉，減少檔案大小，也可以避免外人去看到註解後，進而理解甚至破解你的程式；這種駭客行為，在前端網頁撰寫 JS 很常聽聞。

關於註解撰寫的格式，其實有一些約定俗成的方式。

例如：

工作中：

```
/*
 * 目前進度 ...
 * 2055.0808 Wolke */
```

解釋 class , function , object 的用途

```
/*
 * 參數 f 是 ….
 * 如果是 1，回傳 2
 */
```

版權宣告，通常放在檔頭：

```
/*!
 * Wolke Copyright
 * MIT Licensed
 */
```

其他關於註解的小故事，筆者曾在註解區裡，發現其他公司藏著的徵才訊息等之類奇奇怪怪的內容，偶而留意一下註解，說不定那天你也會在註解區裡，發現什麼有趣的內容喔。

## 1-5-1 註解練習

試練習註解

將下列註解 輸入至編輯區

```
// a one line comment
/*
this is a longer,
multi-line comment
*/
```

## ▌ 1-6 關鍵字 / 保留字

不同的程式語法有各自的關鍵字或是保留字,用途就是寫程式時會用上。

例如在 JS 語法中,有用於宣告的 var let const function class 等;有用於邏輯判斷的 if else 等,或是全域物件的名稱,例如 Date Math 等。

這些被已經被 JS 語法,所保留下來的關鍵字是不能再做他用的,例如不能宣告:

```
let if = 1 ; // 是不行的
let iff = 1 ; // 這可以
```

在之後的章節之中,讀者可以慢慢了解目前在 JS 裡,常用的各式關鍵字 / 保留字。

## ▌ 1-7 變數

變數是暫時存放資料用的,可以將變數想像成是一個箱子,箱子裡面可以是空的,也可以放入一個西瓜,或是 n 個西瓜,當然也可以放入青菜,這個箱子是一個你可以自己決定要放進什麼內容或不放什麼內容的東西。

就算已經放了西瓜,之後你改變主意,也可以將箱子裡的內容物改變,要怎麼使用變數,一切由你決定。

要使用變數功能前,必須撰寫程式來做宣告變數的行為;而什麼是宣告,宣告是指定變數的名稱以及其特性的一種行為。

JS 變數的宣告方式:

```
var a; // 過去
let b; //es5 版之後
```

目前建議盡量用 let 來做宣告;以前 JS 的宣告只有 var ,會造成全域呼叫的問題,雖然還是能用,但目前不那麼建議去用。

```
let a = 1; // a 一開始被宣告為 1
a = '字'; // 變成字串 '字'
a = true; // 變成布林值 true
```

變數的改變，是不是很簡單。

而常數的命名，就是用 const 來做宣告，代表的是唯一值，不變的值。

```
const pi = 3.14; // 宣告圓周率
```

也可以用相同語法在一行裡作連續宣告

```
let a, b = 2, c = true;
let a = b = c = 1; // a, b, c 皆為 1
```

但在程式碼的閱讀上會較不易，故不建議。

## 1-7-1 變數宣告練習

請將下列變數打過並以 console.log 印出

例如：

```
let a = 1;
console.log(a);
let b = 2;
console.log(b);
let c = a + b;
console.log(c);
```

```
---
let x = 5;
let y = 6;
let z = x + y;
let s  = ' 這是一個字串 ';
let s2 = " 這也是一個字串 ";
let s3 = ' 這還是一個字串 ';
let carName;
carName = 'Toyota';
let car1 = 'Toyota', car2 = ' 喜美 ', car3 = ' 保時捷 ';

// 不好的宣告方式  ， 雖然可以，但程式不易讀
const items = getItems(),
    goSportsTeam = true,
    dragonball = 'z';

// 好的宣告方式  ，程式易讀
const items = getItems()
const goSportsTeam = true
const dragonball = 'z'

const a = 10
// 這行程式碼會發生錯誤： "a" is read-only( 只能讀不能寫 )
a = 11
```

## ▌ 1-8 資料型別 Data Type

常聽到 JS 是弱型別語言,那什麼是資料型別,什麼又叫弱型別,那有強型別嗎?

有的。

資料型別、資料類型、資料型態,指的是用來解釋資料的類型,而所謂強型別語言,就是變數一旦被宣告為一種資料型態後,就不能再轉換另一種資料型態。

以上述箱子為例,當該箱子被宣告,只能裝水果時,它就只能換裝西瓜、鳳梨、芭樂,但就不能裝蔬菜或大米什麼的了,這就是強型別;那弱型別,當然就是不管一開始被宣告裝了什麼,都可以再換成別的種類,這就是弱型別,也稱作動態型別。

```
let a = 1; // a 一開始被宣告為數字
a = '字'; // 變成字串
a = true; // 變成布林值
```

JS 的原始資料型別:

- Boolean 布林
- Null 空值
- Undefined 未定義
- Number 數字
- String 字串
- Object 物件

宣告方式:

```
let a = true; // boolean 真值
let b = false; // boolean 假值
let c = null // null 空值
let d ; //Undefined 未定義
let e = 123.567; //Number 數字
let f = ' 字 ' ; //String 字串
let g = {}; //Object 物件
```

檢查型別

```
typeof(a); //boolean 會回傳資料型別
```

## 1-8-1 字串的宣告方式：

```
let text= "word"; //"" 舊的宣告方式
const NAME ='Peter';//'' 舊的宣告方式
let s = 'hello';//'' 新的宣告方式
```

前 2 個都是舊的宣告方式，第 3 個是 es5 後的宣告方式，可以作字串的格式化等，好處多多，建議字串的宣告都改用第 3 個，不過前 2 個還是很常會看到，所以都要知道。

## 1-8-2 資料形別練習

請將下列變數輸入並用 console.log 印出值及用 typeof 印出資料形別

例如：

```
let a = 1;
console.log(a,typeof a);
```

```
const aString = ' 你好 ';
const bString = 'Hello';
const a = 'cat'.charAt(1) ;  //  'a'
const b = 'cat'[1] ;  // 'a'
const aString = 'It\'s ok';
const bString = 'This is a blackslash \\';
const aString = 'hello world';
const aString = 'hello!

 world!';
const intValue = 123;
const floatValue = 10.01;
const negValue= -5.5;
//2 進位
const FLT_SIGNBIT  = 0b10000000000000000000000000000000 ;
const FLT_MANTISSA = 0B00000000001111111111111111111111 ;
//8 進位
const n = 0O755; // 493
const m = 0o644; // 420
//16 進位
const x = 0xFF;
const y = 0xAA33BCv
//boolean
const a = true;
const b = false;
```

## ▌ 1-9 變數命名規則

有關變數的命名，是一門顯學，探討這件事的文章非常的多，每個人的說
法都各有千秋，目前比較主流的慣例就是駝峰命名法，再講解何謂駝峰命
名法之前，筆者兩個經驗先分享：

1. 不要取連自己都容易拚錯的字

2. 盡量一看就懂

### ＊ 不要取連自己都容易拚錯的字

有時候，不知道是為了賣弄文采，還是為了吊吊書袋，很常見到會有人的
變數命名了一些不常見的單字，或者是該說，台灣人比較不熟悉的單字，

但會在國外的開源專案裡看到的單字，就會有台灣人會去用它來做命名，然後程式寫到一半，出現奇怪的 bug 找不到，很多時候僅是因為自己拚錯。

**✲ 盡量一看就懂**

變數名稱盡量一看就懂，不要寫一些很奇怪反向名稱，譬如 cat ，有人會取 notDog ，當然這個舉例是有點誇張，但類似的命名，還真的屢見不鮮。

## 1-9-1 駝峰命名法 Camel-Case

變數名或函式名稱只有一個單字時，都小寫；如果超過 2 個單字時，第一個單字小寫，第二個單字之後，首字母大寫。

例如：

```
let name;
let myName;
let myNickName;
```

就是這麼簡單，增加了程式的識別性與可讀性，也成為目前約定俗成的慣例。

附帶一提，目前常數命名的約定俗成：

```
const PI; // 大寫
const MY_NAME; // 如果有 2 個以上的單字，就用 _ 來連接
const MyName; // 首字母大寫連接起來
```

以上兩種都很常見，看自己喜好，筆者是傾向第一種。

## 1-9-2 變數命名規則練習

將下列內容輸入

```
let numberOfStudents;
var numberOfLegs;
function setBackgroundColor();
class Student{};
const NAMES_LIKE_THIS='Hello';
```

試著自己練習取變數名稱

例如：

```
// 桌子
// 椅子
// 沙發
```

可以將你現在房間內所有物品都試著命名試試 並賦予值後印出

例如：

不變的 門牌號碼

```
const HouseNumber = 118;
const HOUSE_NUMBER = 118;
```

會變的 今天氣溫 , 原子筆數量

```
var temperature =28;
console.log(temperature);
let numberOfPens = 5;
console.log(numberOfPens);
```

等等 越多越好

## ▌ 1-10 陣列 Array

陣列的宣告方式：

```
let person = ['Peter', 23 , true];
```

陣列可以儲存一系列的值。

陣列的取用方式是 zero base，所以是

```
person[0]; // 'Peter'
```

zero-base 的意思，就是以 0 作為第 b1 個索引值，在程式語言中，大部份都是 zero-base。

當然陣列裡面也可以再存陣列

```
let ary = [1,[2,3]];
ary[1][0]; // 2
```

在 JS 裡，沒有定義的陣列位置，也可以再塞值進去，這跟很多語言不一樣。

```
let ary = ['hello']
ary; //['hello']
ary[2]='world';
ary; //['hello', empty, 'world']
```

陣列的用途廣泛，目前還看不出它的好處，但日後寫程式會很常用到。

## 1-10-1 陣列練習

將下列內容輸入並用 console.log 印出

```
const aArray = [];
const bArray = [1, 2, 3];
aArray[0] = 1;
aArray[1] = 2;
aArray[2] = 3;
aArray[2] = 5;
```

多維陣列

```
const magicMatrix = [
    [2, 9, 4],
    [7, 5, 3],
    [6, 1, 8]
];
```

儲存多種資料類型

```
var arr = [1, '1', undefined, true, 'true'];
// 拷貝 (copy) 陣列
const aArray = [1, 2, 3];
const bArray = aArray;

aArray[0] = 100;
console.log(bArray); // 兩個陣列的值都會跟著動，因為其實指向同一個
const aArray = [1, 2, 3];
const copyArray = [...aArray]; // 真正複製出另一個出來
```

# 1-11 運算式＆運算子

要成為運算式至少會有一個 = 的符號

主要有兩種：

1. 用來指派值：

```
let a = 1; //a 被指定為 1
```

2. 包含運算子的運算式：

```
let ans = 2 * 3;
```

除了第一種直接派值之外，其餘的運算式都會含有至少一個以上運算子。

常用的有算術運算子、字串運算子、比較運算子、邏輯運算子，算術運算子、字串運算子說明如下：

## ＊ 算術運算子

| 名稱 | 運算子 | 目的說明 | 範例 | 結果 |
|------|--------|----------|------|------|
| 加法 | + | 兩個數值相加 | 6+3 | 9 |
| 減法 | - | 兩個數值相減 | 6-3 | 3 |
| 乘法 | * | 兩個數值相乘 | 6*3 | 18 |

| 除法 | / | 兩個數值相除 | 6/3 | 2 |
|---|---|---|---|---|
| 遞增 | ++ | 目前值加 1 | let i=6;<br>i++; | 7 |
| 遞減 | -- | 目前值減 1 | let i=6;<br>i--; | 5 |
| 取餘數 | % | 兩個數值相除後回傳餘數 | 6 % 3 | 0 |

## ＊ 字串運算子

用於連接兩邊的字串資料，有 ' ' " " ` ` 等撰寫方式

```
let n = ' 這是 ' + ' 字串 ';
```

也可以連接變數：

```
let n = ' 彼德 ';
console.log(' 我是 ' + n );
```

不過在 ` ` 問世之後，都被改成：

```
let n = ' 彼德 ';
console.log(` 我是 ${ n }`);
```

字串的格式化較為直覺且易讀。

至於比較運算子、邏輯運算子會在決策與迴圈章節再做說明。

## 1-11-1 運算式 & 運算子練習

試著計算下列數學題目並印出答案

例如：

一雙襪子有 2 隻，5 雙共有幾隻襪子？

```
let socks = 2 ;
let ans = 2 * 5;
console.log(ans);
```

題目：

1.　一串香蕉有 5 根，9 串香蕉有幾根？

2.　存錢筒裡原來有 38 元，美琳昨天 存入 18 元，今天再存入 22 元，撲滿裡現在一共有多少元？

3.　小智的體重是 33 公斤，他比弟弟 重 12 公斤，小智和弟弟的體重合起來是幾公斤？

4.　每 5 個饅頭裝成 1 袋，現在有 35 個饅頭，全部 裝完，可以裝成幾袋？

5.　有 39 人玩遊戲，男生有 22 人，每 5 人分成 1 組，男生最多可分成幾組？還剩下幾人？

6.　小花買了一塊長 5 公分的橡皮擦，用掉 16 毫米 後，還剩下幾毫米？

7.　5 個一、9 個十和 4 個千合起來 是多少？

8.　資源回收場有 1803 個鐵罐和 98 個玻 璃罐，這兩種罐子相差幾個？

9.　有一幅拼圖，拼了 1468 片，剩下 632 片還沒拼完，這幅拼圖有幾片？

10.　鮮奶一瓶賣 38 元，炭治郎買了 9 瓶，禰豆子買 4 瓶，炭治郎比禰豆子多付多少元？

11.　旺達冒險工廠生產 998 顆巧克力，每 36 顆裝成一 盒，共可裝成幾盒？

還剩下幾顆？

12. 一台相機 1200 元，一包相片紙 230 元。阿翰買了 3 台相機和 16 包相片紙後，還剩下 55 元，請問阿 翰原本有幾元？

13. 爺爺買了 4 臺冷氣，共付了 200 張 500 元 、30 張 1000 元、5 個 100 元、5 個 10 元 和 9 個 1 元，請問這 4 臺冷氣總共是多少 元？

# 2

## IDE 篇

## 2-1 學習目標／演練成果

學習 IDE 的基礎

## 2-2 IDE 是什麼？

工欲善其事，必先利其器。

所謂 IDE 就是 Integrated Development Environment 的縮寫，是一種輔助程式開發人員撰寫程式的利器。

筆者最早的接觸 IDE 應該算是 linux 系統的 vi，但 vi 其實以今日的標準來說，應該不算 IDE，算是小作家。

目前 IDE 的選擇，其實是因應著開發上的需求，而有所不同，例如開發 Android APP，最早是用 JAVA，所以要用 Eclipse，後來用 Android Studio，開發 iphone APP，撰寫的是 Object-C 所以用 xcode …

不同的開發目地，會有不同 IDE 的選擇，但目前的 IDE 的主流配置，有些共通點，例如得有語言編輯區，語言編輯區得有程式碼提示的功能，甚至漸漸的還有用 AI 去做程式碼補上的功能；並且有命令區塊可以下命令。

## 2-3 codesandbox 可以做什麼？不能做什麼？

很多人對於用 codesandbox 來做初學者教學是嗤之以鼻的；所以我得在這裡做一個為什麼用 codesandbox 的說明：

### 2-3-1 codesandbox 的好處是什麼：

1. 免安裝，隨開隨用
2. 幾乎免費，無開發上限制

#### ＊ 免安裝，隨開隨用

相對於 VS Code 之類的 IDE，需要做安裝的動作，而安裝這個動作，在筆者的經驗裡，一百台電腦或有一百種狀況，尤其是以開發 node.js 的環境來說，所需要安裝的工具還頗多的，對於初學程式來說，還沒有開始撰寫到一行 code ，可能就會先被安裝環境給打敗了。

### ＊ 幾乎免費，無開發上限制

相較於其他競品，例如 gitpod 在過了 50 小時的試用之後，就得付費；或 StackBlitz 有開發上限制，用的是 PWA 技術，導致無法開對外 port，也就無法先將其作為後端應用的開發；而其他的競品，常見的就是專案數量上的限制等。

## 2-3-2 codesandbox 的缺點：

1. 程式碼開放
2. 不能設斷點

### ＊ 程式碼開放

codesandbox 木著開放精神，在免費模式下，你所打的每一行 code 都是公開的，但也有提供 Secret key 的功能，讓你可以放較機密的內容，例如：金鑰等，也就不怕被外洩出去了。

### ＊ 不能設斷點

在開發程式專案上，對於程式設計師來說，有個很重要的功能，就是設斷點，不幸的是 codesandbox 目前還不能設斷點。

但設斷點的主要目的是找出那裡錯誤，通常在大型專案，協同工作時，會較容易發生程式不如預期的結果，此時就得設斷點，去尋找出是那個環節出了錯誤。

但對於新手來說，其實 console.log 就很夠了。

為什麼呢，因為初學的程式行數其實很少，不超過 50 行，所以設斷點這件事也就沒那麼必要了，而且初學者先不要依賴斷點工具，培養對程式撰寫的熟悉度，會是比較好的。

## ▌ 2-4　codesandbox 設定

我們寫程式，首先要有個編輯器 IDE：Integrated Development Environment 整合開發環境，編輯器的作用就是方便我們撰寫程式，其實就算用小作家來寫都可以，不過小作家非常的不方便，因為我們的人腦有限，沒辦法去記住這麼多的指令。

所以我們就需要編輯器來提示我們，撰寫程式時所需要的一些關鍵字，或是之前給你自己定義的一些參數。

這本書裡所採用的是 codesandbox 線上編輯器 ，好處是：

1.　不需要再做安裝的動作

2.　自動架設好連外網的伺服器

3.　可以直接引用 github 的開源專案

首先要使用 codesandbox 非常簡單，打開 google 搜尋 codesandbox，code sand box 的意思 就是程式的沙盒。

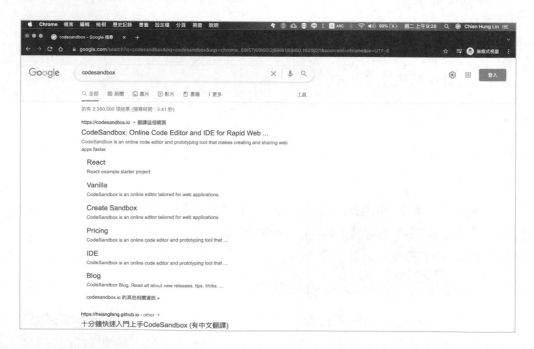

第一個就是了，點進去

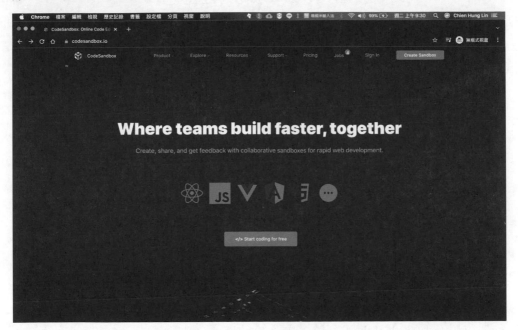

在開始使用之前，還是要提醒各位，codesandbox 的免費版本會將所有程式碼公開，所以：

密碼什麼的，千萬不要明碼的放上去，

密碼什麼的，千萬不要明碼的放上去，

密碼什麼的，千萬不要明碼的放上去，

重要的事情說 3 次，請注意。

點擊右上角的 signin

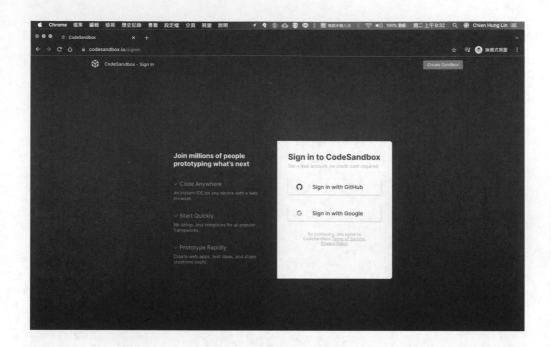

這裡可以看到 codesandbox 的 3 個訴求：

1. Code Anywhere：只要有網路，在任何地方，都可以連進 codesandbox
   撰寫你的專案

2. Start Quickly：不需要任何安裝的前置作業，就可以開啟你所需要編輯
   環境

3. Prototype Rapid：創作你的網站程式，測試想法可行性，都非常的容易

目前支援兩家賬號的第三方 sign in：

1. github：有很多的開源程式，都會在 github 上面，以前它是自己經營的，
   收費比較貴，後來被微軟給收購，所以他現在免費的限度都提高了，
   建議所有要學程式的朋友，都應該開一個 github 帳號，日後會有用處。

2. Google Account ：就是用 Google 的帳號登入。

不管選擇那種方式，都可以 sign in 進來使用 codesandbox 的服務。

進來之後，就點 Create Sandbox

可以看到許多知名的 framework 已經被 codesandbox 引用成為 template，沒看過這些也沒關係，在初學來說，我們先嘗試開啟一個最基礎的套件 Node HTTP Server 。

點下去之後， codesandbox 便會幫我們將環境給開好

Node HTTP Server 的 IDE，初始狀態，主要分為幾個區域：

1. 工具列：各種設定

2. 檔案目錄：檔案開啟的操作

3. 已安裝套件：npm 新增安裝或移除

4. 其他資源

5. 程式碼編輯區：檔案開啟後的程式碼撰寫，開啟圖檔，則會出現圖片

6. 網頁預覽區：可以看到編輯中的網站在 browser 瀏覽器下的樣子

7. Terminal 命令工具：可以下命令

出現這個畫面，代表著 codesandbox 已經幫各位開好了一個最基本的 node. js 網站伺服器了，而且你還可以開始撰寫程式碼。

## ▌ 2-5 package.json 簡介

```
{} package.json  ×

    {
      "name": "refetch-interval",
      "version": "1.0.0",
      "main": "index.js",
      "license": "MIT",
      "dependencies": {},
      "scripts": {
        "start": "node index.js"
      },
      "devDependencies": {
        "@types/node": "^17.0.21"
      }
    }
 14  |
```

package.json 就是 node.js 的設定檔。

1. name：就是專案名稱

2. vesion：是版本號

3. main：是程式的主要進入點

4. license：是版本宣告

5. dependencies：是相依的套件

6. scripts：是可以執行的命令，start 是起始的指令，通常我們在 terminal 裡，下 npm start 就會去執行 npm start 裡的命令，而 codesandbox 的預設，一開啟就會執行 npm start

7. devDependencies：是開發時所相依的套件，也就是 release 時，用不到的套件，例如這裡的 "@types/node": "^17.0.21" 是用在撰寫程式時，提示說明程式碼用的，自然在 release 時用不到，不用被編譯進去。

## 2-5-1 安裝第一個套件 nodemon

npm 就是 node package manager 的縮寫，意思就是 node.js 的套件管理器；
在越來越開放的今天，node.js 之可以蓬勃發展，其中一個原因，要歸功於
node.js 社群不斷的有新的 feature 的套件誕生，方便開發者們去引用，做為
開發之用。

而在 node.js 裡，套件的撰寫、安裝及管理是相當簡單的，首先來安裝我們
的第一個套件 nodemon，這是一個方便我們撰寫程式所使用的 Tool 工具。

codesandbox 支援兩種方式安裝。

**\* 透過 Add dependency 安裝套件**

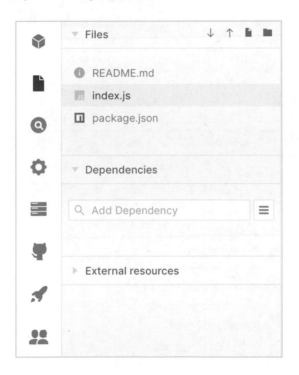

1. 在 codesandbox，直接在 Dependencies 下的 Add dependency打上
   nodemon

2. 點擊 nodemon

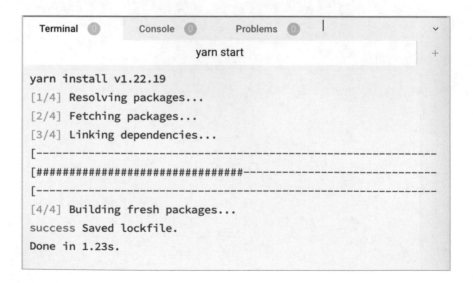

3. 可以看到命令工具 terminal 裡，codesandbox 自動幫我們安裝。

\* **透過 對話框 安裝**

1. 點 ▤ 後

2. 跳出對話框後，直接輸入套件名稱，可以看到詳細的說明

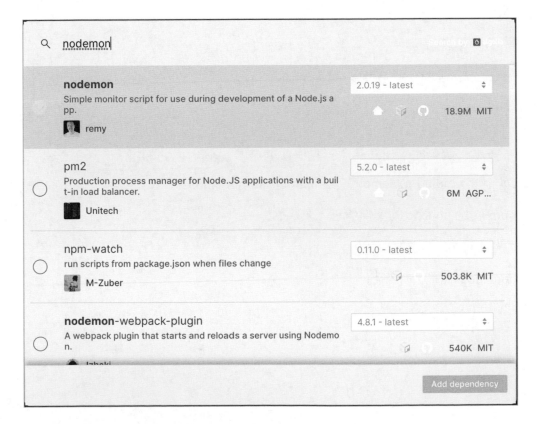

3. 選取要安裝的套件，點擊 Add dependency ，便可以安裝進去。

## * 透過 terminal 安裝

也可以用傳統的方式,直接透過 terminal 下指令。

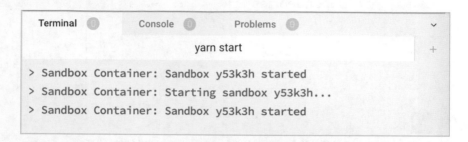

1. 點擊右邊的 + 號

會多了一個新的 terminal ,再點一次就會多一個,以此類推。

2. 輸入 npm i nodemon 或是 npm install nodemon

npm 就是 node.js 的套件管理工具

i 就是 install 安裝的意思

nodemon 就是套件名稱

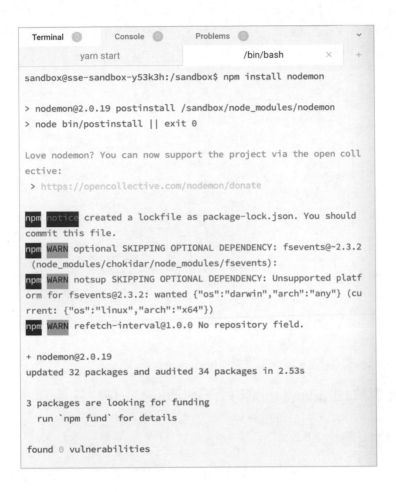

然後就可以看到安裝開始到完成

在 Dependencies 這裡就會看到套件的名稱了。

也可以開啟 package.json

在 dependencies 項裡，就可以看到 nodemon 了。

附帶一提，Add dependency 所下的指令 yarn 是 fb 所另外開發的。

## 2-5-2  變更版本號

在  點擊後，就可以選擇其他版本作安裝，在有些程式不向下相容時，非常有用。

### 2-5-3 移除套件 nodemon

點 x 就可以直接移除套件了。

## ▌ 2-6 執行 nodemon

使用 nodemon 的好處，就是 nodemon 會幫我們監控檔案有沒有變動，一有變動會重啟程式，而我們不需要中斷指令，再重下指令。

1. 首先至 package.json 修改 start 的參數為 nodemon index.js，存檔。

```
"scripts": {
  "start": "nodemon index.js"
},
```

2. 存檔後，至 Server Control Panel 選 Restart Sandbox

3. 可以看到 terminal 有重啟了，改成一開始下的指令是 nodemon index.js

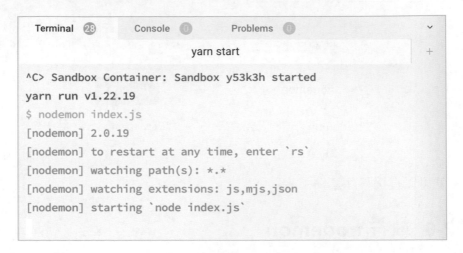

4. 可以去修改 index.js "Hello World" => "Hello Baby"

```
var http = require("http");

//create a server object:
http
 .createServer(function (req, res) {
   res.write("Hello Baby"); //write a response to the client
   res.end(); //end the response
 })
 .listen(8080); //the server object listens on port 8080
```

5. 存檔後，可以看到 terminal 自動重啟

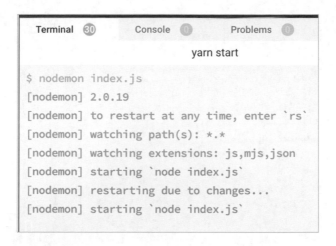

6. 重新整理 Browser

字串就跟著變了

另外 package.json 裡的 dependencies 指的是必要的相依元件。

而 devDependencies 是開發時才會用到的元件。

如果 nodemon 被裝到了 dependencies，我們就應該編輯 package.json ， 將 nodemon 套件從 dependencies 移至 devDependencies ，避免日後專案發佈版本時，將只是協助開發用的工具 nodemon 給包裹進去，造成專案的肥大。

package.json:

```
{
  "name": "xxx",
  "version": "1.0.0",
  "main": "index.js",
  "license": "MIT",
  "dependencies": {},
  "scripts": {
    "start": "nodemon index.js"
  },
  "devDependencies": {
    "@types/node": "^17.0.21",
    "nodemon": "2.0.16"
  }
}
```

## ▌ 2-7 codesandbox 的狀況排除

### 2-7-1 無法正常運行

早年 codesandbox 剛問世時，online IDE 的觀念受到大家的歡迎，但很多人用之後，也隨之出現很多的問題，例如：網路延遲，機器的資源開不了新的 container，或是一些大大小小的靈異狀況，但經過多年的運營，其實已經改善很多；可要是在撰寫程式時，你發現了以下狀況，可以試著使用如下方式去排除。

＊ 狀況

1. 瀏覽器畫面跳 error
2. 更新 檔案內容 後，瀏覽器畫面沒有變化

＊ 常見盲點

1. 沒有按 save
2. 參數名稱拚錯

* **重啟方法 1**

1. 瀏覽器直接按 refresh

* **重啟方法 2**

1. 到 Server Control Panel 按 Restart Sandbox

* **重啟方法 3**

1. 到 Server Control Panel 按 Restart Server

* **重啟方法 4**

1. 按 fork

* **重啟方法 5**

1. npm i nodemon

2. 改 package.json

3. start : nodemon index.js

4. 按 fork

* **重啟方法 6**

bug: 例如命名 app.js 並以 app.js 為進入點，未修正 package.json 的 main 為 app.js ，會導致無法運行，改一下之後，要 restart sandbox ，可能是 codesandbox 的預設進入點都是 index.js，所以改進入點或造成錯誤

## 2-7-2 IDE 練習

IDE 就是編輯程式所不可或缺的工具軟體

試開啟其他的 template 練習之

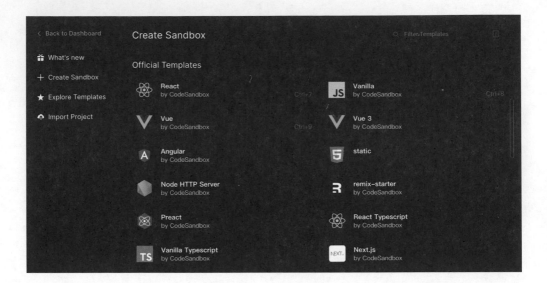

# 3

## 程式學習篇
## 函式方法與物件

## ▌ 3-1 開發環境

1. chrome 瀏覽器內建的主控台 console
2. codesandbox 開啟 node http server 的 sandbox

## ▌ 3-2 學習目標／演練成果

藉由實作範例，來理解函式及物件

1. 函式：攝式華式的換算
2. 物件：各式物件的練習

## ▌ 3-3 函式 & 箭頭函式

函式讓你可以將一系列的程式碼，藉由括號 { } 包裹起來，以完成一項特定的工作；在實務上通常會將重複執行的工作，宣告成一個函式，進而來做之後的呼叫，好處是減少程式碼，且增加程式的易讀性和維護性。

### 3-3-1 宣告及呼叫

這邊可先利用 chrome 的 console 來做練習

#### ＊ 傳統函式宣告

```javascript
function sayHello() { //sayHello 就是函式的名稱
     return 'hello';
};
```

```javascript
const sayHello = function(){
   return 'hello';
};
```

return 就是將後面的值傳出去，也就是回傳值，當然你也可以不 return 值出去。

* 箭頭函式宣告

```
let sayHello = ()=>{
    return 'hello';
};
```

es5 版之後，增加了箭頭函式，也可作為函式的宣告。

從此函式的宣告，你可以選擇省略掉 function 這個關鍵字，使程式碼上看起來像是變數的宣告方式，進而增加了程式的易讀性。

* 箭頭函式宣告可以省略掉 { } 跟 return

```
let sayHello = ()=> 'hello';
```

{ 直接接著 return ，還可以再省略掉 return 跟 { }，依舊可以將值傳出去，不過初學可能會有點混亂，先能看懂這個就好；等程式碼更熟悉了之後，再做這樣子的撰寫。

是認真講的，先不要假勞，看過太多這樣的 bug 。不過本書大部份的範例還是會使用箭頭函式，因為比較美觀易讀。

* 回傳值：任何的資料型態都是可以作為回傳值的

```
let getFunction = ()=>{
    return ()=>{console.log(' 我是函式 ')};
};
getFunction()(); // 呼叫 getFunction() 後，拿到一個函式，所以必須再呼叫 () 去
執行它，看起來有點怪怪的，但在 JS 語法裡，這沒有錯。
```

* 呼叫函式

```
let text = sayHello();
console.log(text);  // text = 'hello';
```

函式宣告完成之後，就可以直接呼叫了。

## \* 函式帶參數的宣告

以攝式華式換算為例

```
// 傳統
function getF(c){ //c 就是 getF 的參數
        return c * 9 / 5 + 32; // 可以直接引用 c
};

function getC(f){
        return (f -32) * 5 / 9 ;
};

// 新式
let getF = (c)=> c * 9 / 5 + 32;

let getC = (f)=> (f -32) * 5 / 9;
```

## \* 呼叫函式

```
getC(212); // 100
getF(0); // 32
```

這裡就可以看出將程式碼包成函式的好處之一，就是減少程式碼的撰寫。

## \* 指定預設值的函式宣告

es6 之後，可以加上參數預設值

```
let getF = (c=0)=> c * 9 / 5 + 32; //es6
getF(); // 32
```

也可直接用 … 將所有參數轉成陣列

```
let getPlus = (...p)=>{ // 參數變成一串陣列
    return p.reduce(
        (a,b)=>{
            return a + b
        }
    )
}
getPlus(1,2,3,4); //10
```

reduce 是陣列物件的內建函式，後面會再講到 reduce。

這裡先寫出來，是因為通常用 … 這個來作函式的參數時，裡面會用陣列物件的內建函式來做程式的撰寫。

## 3-3-2 解釋 node http server 的函數

理解了函式及參數之後，至 codesandbox 開啟 node http server 的 sandbox，裡面的 index.js 為：

```
var http = require("http");

//create a server object:
http
 .createServer(function(req, res) {
    res.write("Hello World!"); //write a response to the client
    res.end(); //end the response
 })
 .listen(8080); //the server object listens on port 8080
```

裡面的

```
function(req, res) {
    res.write("Hello World!"); //write a response to the client
    res.end(); //end the response
}
```

也是函式。

而 http.createServer() 是物件的內建函式，下一節會再解釋何為物件。

這裡可以看到 function(req, res){...} 函式，被當作參數傳進去 createServer 這個函式裡，所以 JS 裡的參數是任何的資料型別，都是可以當作參數的。

而這裡會覺得奇怪的是，req 跟 res 是什麼時候傳值進去的，為什麼可以有 res.write() 的呼叫？

其實是 http.createServer() 去做了 function 的呼叫並且傳值進去，這在 JS 裡是很常見的程式寫法。

可以解釋成：

```
let a = (f)=>{ // a 的參數 f 預期就是函式，
    let n = 1;
    f(n); // f 傳入 a 後，被執行時，傳了 n 做為參數。
}
a((i)=>{ // 函式傳入 a ，被倒入 n 也就是值 1
    console.log(i+1);
}); // 2
```

### 3-3-3 函式練習

在 chrome console 輸入下列內容並印出值

ps: chrome console 的換行是 Ctrl + Enter

```javascript
const addOne = function(value){
    return value + 1
}

const addOneAndTwo = function(value, fn){
    return fn(value) + 2
}
 const addOne = function(value){
    return value + 1
}

const addOneAndTwo = function(value, fn){
    return fn(value) + 2
}

addOneAndTwo(10, addOne)
function sum() {
  return arguments[0]+arguments[1]
}

console.log(sum(1, 100))
function sum(x, y, z) {
  return x+y+z
}


console.log(sum(1, 2, 3))
console.log(sum(1, 2))
console.log(sum(1, 2, 3, 4))
console.log(sum('1', '2', '3'))
console.log(sum('1', '2'))
console.log(sum('1', '2', '3', '4'))
function sum(...value) {
    let total = 0
    for (let i = 0 ; i< value.length; i++){
        total += value[i]
    }
    return total
}
```

```javascript
// 上面 console.log 的內容 再打一次試試
function addOuter(a, b) {

    function addInner() {
        return a + b
    }

    return addInner()
}

addOuter(1, 2)
function addOuter(a, b) {
    return addInner(a, b)
}

function addInner(a, b) {
    return a + b
}

addOuter(1, 2)

// 匿名函式
(function(){
    console.log('IIFE test1')
}())

function test2(){
    (function(){
        console.log('IIFE test2')
    }())
}

test2()

// 將上述改成箭頭函式 再實作一次
```

## ▌ 3-4 變數作用範圍 / 變數作用域 scope

### 3-4-1 let 宣告

理解了函式及參數之後，回到 codesandbox 的 node http server 的 index.js

```
var http = require("http");

//create a server object:
http
 .createServer(function(req, res) {
   res.write("Hello World!"); //write a response to the client
   res.end(); //end the response
 })
 .listen(8080); //the server object listens on port 8080
```

在第二行宣告

```
let text = 'outside'; //global scope
```

將

```
res.write("Hello World!");
```

改成

```
res.write(text);
```

所以程式碼變成是：

```
var http = require("http");
let text = 'outside';//global scope
//create a server object:
http
 .createServer(function(req, res) {
   res.write(text); //write a response to the client
   res.end(); //end the response
 })
 .listen(8080); //the server object listens on port 8080
```

重新整理右邊的 browser 會得到 'outside'

當 res.write(text); 在執行時，在函式作用範圍內，找不到 text 時，就會去尋找 text。

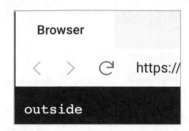

再加一行到 function(req,res)... 裡

```
let text = 'middle';
```

所以程式碼變成是

```
var http = require("http");
let text = 'outside';
//create a server object:
http
 .createServer(function(req, res) {
   let text = 'middle';
   res.write(text); //write a response to the client
   res.end(); //end the response
 })
 .listen(8080); //the server object listens on port 8080
```

重新整理右邊的 browser 會得到 'middle'

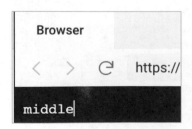

我們再加一個函式 f 進去，

```
let f = () => {
    let text = "inside";
    res.write(text); //write a response to the client
    res.end(); //end the response
};
```

程式碼變成

```
var http = require("http");
let text = 'outside';
//create a server object:
http
 .createServer(function (req, res) {
  let text = "middle";
  let f = () => {
    let text = "inside";
    res.write(text); //write a response to the client
    res.end(); //end the response
  };
  f();
})
 .listen(8080); //the server object listens on port 8080
```

重新整理右邊的 browser 會得到 'inside'

因為函式 f 的作用域裡面已經有一個 text 了，它就不會向外去找 text 了。

由此可知，變數被呼叫的順序的邏輯，是由函式內部開始尋找變數，一層一層往外找出去。

這個邏輯關係，在程式裡叫做變數作用範圍

而 JavaScript 有兩種範圍：

- 全域變數：在函式作用域 (function scope) 之外宣告的變數，在整個程式中都可以被存取與修改，例如前述的範例的 let text = 'outside';

- 區域變數：在函式作用域 (function scope) 內宣告，每次執行函式時，才會建立的區域變數，而這個函式之外的所有程式碼都不能存取這個變數，例如前述的範例的 let text = 'inside';

但是 var 宣告跟 let 宣告在 JS 裡又不太一樣了。

加個 if 在 let text = 'inside'; 的外面，程式碼改成

```javascript
var http = require("http");
let text = 'outside';
//create a server object:
http
 .createServer(function (req, res) {
   let text = "middle";
   let f = () => {
     if (true) {
       let text = 'inside';
     }
     res.write(text); //write a response to the client
     res.end(); //end the response
   };
   f();
 })
 .listen(8080); //the server object listens on port 8080
```

重新整理右邊的 browser 會得到 'middle'

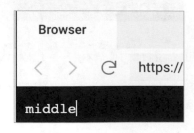

可以理解為 let text = 'inside' 已經在 if 作用域內了，所以 res.write(text);
只能往外圍尋找到 let text = 'middle';。

但改成 var text = 'inside';

程式碼變成

```
var http = require("http");
let text = 'outside';
//create a server object:
http
 .createServer(function (req, res) {
   let text = "middle";
   let f = () => {
     if (true) {
       var text = 'inside';
     }
     res.write(text); //write a response to the client
     res.end(); //end the response
   };
   f();
 })
 .listen(8080); //the server object listens on port 8080
```

重新整理右邊的 browser 會得到卻是 'inside'

這是因為

● let 與 const 是區塊作用域 (block scope)

● var 是函式作用域 (function scope)

也就是 var text = 'inside' 在函式 f 內都可以被存取，但是 let 只存在於 {
... } 區塊內，這點要注意。

這裡之所以要提變數作用範圍 scope，是因為在實務上，很常見到這類的 bug，當程式碼逐漸增加時，程式設計師 programer 可能會忘了曾經在作用範圍之外，命名過相同的變數，而導致產生不如預期的結果。

為了避免發生這樣的錯誤，目前有幾個原則可以遵守：

1.  最外圍只命名常數

2.  避免一圈又一圈的函式

3.  一個 scope 內不要做太多的事，能拆就拆

4.  不要用 var 作宣告

### 3-4-2 函式的變數作用範圍練習

輸出下列內容 並印出值

```
if (true) {
  var x = 5
}

console.log(x)
if (true) {
  let y = 5
}

console.log(y)
```

## 3-5 物件

由物件導向程式設計而來；曾經 ( 差不多 40 年前吧 ) 的程式撰寫只到函式就結束了，而物件導向程式設計的發明，將程式撰寫抽象化，簡單的說法，就是將程式的撰寫用物品比擬之類。

這邊可先利用 chrome 的 console 來做練習

## 3-5-1 宣告方式

```
let obj = {};
```

## 3-5-2 物件在實務上的使用為：

* **1. 比擬為物品**

```
let cat = {
        name : '加非',
        month:10,
        sex: false
};
```

name , month , sex 通常被稱為屬性或特性。

* **2. 被用來做函式的集成**

```
let excise = {
        run : ()=>{ console.log('跑步中')},
        meow:  ()=>{ console.log('喵喵叫')},
};
```

當被用作函式的集成時，通常會被包成 module 模組化，例如 node.js 的內建 http 套件

* **3. 上述兩種都用**

```
let cat = {
        name : '加非',
        month:10,
        sex: false,
        run : ()=>{ console.log('跑步中')},
        meow:  ()=>{ console.log('喵喵叫')},
};
```

### 3-5-3 呼叫方式

```
cat.name; //' 加非 '
cat.meow(); //' 喵喵叫 '
```

或是像陣列一樣使用 [ ] ，只是裡面要用參數名稱

```
cat['month']; //10
```

\* **變更屬性值**

```
cat.name = ' 歐弟 ' ;
```

\* **新增新的屬性**

```
cat.age = 1;
```

\* **移除屬性**

```
delete cat.age;
cat.age; // 再呼叫取用就變成 undefined
```

\* **取出並複製屬性**

```
let {name , meow} = cat;
name; //' 加非 '
meow(); //' 喵喵叫 '
```

\* **簡寫物件屬性宣告**

```
let name = ' 加非 ';
let meow = ()=>{ console.log(' 喵喵叫 ')};

let cat = { // 直接用變數名稱，宣告物件時，就會直接被代入
    name,
    meow
};
cat.name; //' 加非 '
cat.meow(); //' 喵喵叫 '
```

## 3-6 物件與類別

可以將類別視為宣告一個物件的模版，並用此模版來產生各個不同的物件，
但會擁有相同的屬性。

### 3-6-1 類別的宣告方式

```
function Cat(name){ // 傳統
  this.name = name;
  this.run = ()=>{ console.log(' 跑步中 ')},
}

class Cat{ // 正式
  constructor(name){// 建構函式
    this.name = name;
  }
  run(){ console.log(' 跑步中 ')}
}
```

\* 使用方式

```
let kittyCat = new Cat('kitty'); // 產生一個叫做 kittyCat 的物件
let pinkyCat = new Cat('pinky'); // 產生一個叫做 pinkyCat 的物件
```

使用類別新增實體物件，不管新增幾個新的實體，都不需要再去重新定義
內部的屬性等。

* 呼叫方式

```
kittyCat.name; //'kitty'
kittyCat.run(); //'kitty 跑步中 '
```

* 常見的用法

```javascript
class Cat {
  setName(name) {
    this.name = name;
    return this; // 將自己傳出去
  }
  run() {
    console.log('${this.name} running');
    return this; // 將自己傳出去
  }
  meow() {
    console.log('${this.name} meowing');
    return this; // 將自己傳出去
  }
}

new Cat().setName('peter').run().meow();
// peter running
// peter meowing
```

這也是目前在 JS 撰寫上很常見的技巧。

## 3-6-2 撰寫溫度換算 API

我們現在大約了解了物件與函式，我們來試著替 node http server 新增一個華式換算攝式的功能

1.    至 codesandbgox 新增 node http server 的 sandbox

2.    我們引用 url , querystring 兩個內建模組

```
const url = require('url'); // 引用內建模組 url
const querystring = require('querystring');// 引用內建模組 querystring
```

3. 並且加上

```
const getF = (c=0)=> c * 9 / 5 + 32;
```

4. 將 function 改寫成

```
function (req, res) {
   let { query } = url.parse(req.url); // 利用 url 的內建函式，得到一個物件，
需要 query 值  / ? c=100
   console.log(query);
   let { c } = querystring.parse(query); // 得到 c 的值
   res.write("" + getF(c)); // 加 ""+ 目地是將後面的回傳值是數字型態轉換為字
串， write 函式只能寫入 字串 資料型態
   res.end(); //end the response
}
```

全部程式碼為：

```
var http = require("http");
const url = require('url');
const querystring = require('querystring');

const getF = (c=0)=> c * 9 / 5 + 32;
//create a server object:
http
 .createServer(function (req, res) {
    let { query } = url.parse(req.url);
    console.log(query);
    let { c } = querystring.parse(query);
    res.write("" + getF(c));
    res.end();
})
 .listen(8080); //the server object listens on port 8080
```

在 browser 欄位根目錄後輸入 /?c=100，就會在 browser 的內容區裡得到
212：

這就是一個很簡單的 API 應用；你可以將網址丟給朋友玩玩看

```
https://xxx.sse.codesandbox.io/?c=50
```

## 3-6-3 物件練習

請將下列內容輸入並印出

```
const emptyObject = {};

const player = {
    fullName: 'Inori',
    age: 16,
    gender: 'girl',
    hairColor: 'pink'
};

const aArray = [];
const aObject = {};

const bArray = ['foo', 'bar'];
const bObject = {
    firstKey: 'foo',
    secondKey: 'bar'
};

bArray[2] = 'yes';
bObject.thirdKey = 'yes';
```

```
console.log(bArray[2]); //yes
console.log(bObject.thirdKey); //yes
const luke = {
  jedi: true,
  age: 28,
};

function getProp(prop) {
  return luke[prop]
};
const isJedi = getProp('jedi');
var person = {
  firstName: "John",
  lastName : "Doe",
  id       : 5566,
  fullName : function() {
    return this.firstName + " " + this.lastName;
  }
};
```

## ▌ 3-7 全域物件

所謂全域物件就是在程式的任何角落都可以被呼叫的物件。

例如在前端 JS，就是 windows , document 等物件，在後端 node.js 環境，則是 process 等。

呼叫這類全域物件，只需要打上變數名稱即可引用呼叫。

```
process.env // 引用環境變數
```

但事實上，在 JS 裡，可以將萬物皆看成物件。

例如宣告完成的變數，可以依其資料型態，去呼叫其資料型態所內建的函式或屬性。

例如

```
let e = 123.567; // e 是 Number 數字資料型態
e.toFixed(0); // 便可以呼叫 Number 的內建函式
```

### 3-7-1 關於 this

原則上，程式撰寫物件時， this 是用來指向物件內部的值用的。

但這僅只於原則上。

例如：

```
let obj = {
    v:1,
    getV1:()=>this.v,// 箭頭函式
    getV2:function (){ return this.v}, // 一般函式
}
obj.getV1(); //undefined
obj.getV2(); //1
```

可以看到呼叫 obj 所定義的 getV1 和 getV2 函式，明明都是函式，卻因為使用不同的函式宣告方式，而得到不同的結果。

但如果是使用類別來作宣告

```
class Obj{
    constructor(v){
        this.v=v;
        getV1=()=>this.v; // 箭頭函式的宣告
    }
    getV2(){ return this.v} // 一般函式
}
let obj = new Obj(1);
obj.getV1(); //1
obj.getV2(); //1
```

兩者則可以得到預期的結果。

故在 this 的使用上，必須小心謹慎，避免產生邏輯上，不如預期的錯誤。

## 3-8 全域物件 STRING 字串

String 為基礎資料型態

### 3-8-1 宣告 String 物件

```
let text = 'Hello World';
```

#### ＊ 常用屬性及內建函式

```
// 取得字串中的字元數量
text.length; //5
// 取得該字元第一次出現的位置
text.indexOf('e'); //1 因為是 zero-base，所以是 1
// 指定一字元分割字串為陣列
text.split(' '); //['Hello' , 'World']
// 取得兩索引值之間的字串
text.substring(1,3); // 'el'
// 取代前值
text.replace('World','Baby'); // 'Hello Baby'
```

要注意的是 String 物件的函式並不會去改變原本的值

```
text; // 'Hello World'
```

所以如果想要改變原始的值，記得要 = 接回去

```
text = text.replace('World','Baby');
text;// 'Hello Baby'
```

#### ＊ 格式化

在只有 "" 跟 " 時代裡，要做字串的排版，只能用 +

```
let h =  "hello";
let w =  "world";
let t = h + " " + w + "!"; //需要加一個空白
t; // 'hello world'
```

你可以想像當變數一多起來之後，程式碼會有多亂

es5 之後 " 問世之後，裡面可以加參數，格式化也就輕鬆多了

```
let h =  "hello";
let w =  "world";
let t = '${h} ${w}!'; // 直接使用 ${} 代入參數，變得乾淨整齊
t; // 'hello world!'
```

而 '${}' 不是只能放參數，還可以放運算式

```
'1 + 1 = ${ 1 + 1 }'; // 1 + 1 = 2
```

也可以運行函式

```
let n = 3;
let f = (i)=>i+1;
'No. ${f(n)}';  // 'No. 4'
```

## 3-8-2 全域物件 STRING 練習

輸入下面內容並印出值

```
例 1 :
var txt = "ABCDEFGHIJKLMNOPQRSTUVWXYZ";
var sln = txt.length;
印出 console.log(sln);
例 2 :
var str = "Please locate where 'locate' occurs!";
var pos = str.search("locate");
印出 console.log(pos);


var str = "Please locate where 'locate' occurs!";
var pos = str.indexOf("locate");


var str = "Please locate where 'locate' occurs!";
var pos = str.lastIndexOf("locate");


var str = "Apple, Banana, Kiwi";
var res = str.slice(7, 13);


var str = "Apple, Banana, Kiwi";
var res = str.slice(-12, -6);


var str = "Apple, Banana, Kiwi";
var res = str.slice(7);


var str = "Apple, Banana, Kiwi";
var res = str.slice(-12);


var str = "Apple, Banana, Kiwi";
var res = str.substring(7, 13);
```

```
var str = "Apple, Banana, Kiwi";
var res = str.substr(7, 6);

var str = "Apple, Banana, Kiwi";
var res = str.substr(7);

var str = "Apple, Banana, Kiwi";
var res = str.substr(-4);

var str = "Please visit 1990!";
var n = str.replace("1990", "1989");

var text = "Hello" + " " + "World!";
var text = "Hello".concat(" ", "World!");

var str = "    Hello World!
";
alert(str.trim());

var str = "HELLO WORLD";
str.charAt(0);

var str = "HELLO WORLD";
str.charCodeAt(0); // 會傳回 H 的字元編碼

var str = "HELLO WORLD";
str[0];

var str = "HELLO WORLD";
str[0] = "A";              // 可以給但不會有作用
str[0];                    // 傳回 H

var txt = "a,b,c,d,e";
txt.split(",");
```

## ▌ 3-9　全域物件 NUMBER

Number 數值為基礎資料型態，你可能會聽過在強型別的語言裡，例如 JAVA C++ ⋯ 會將數值的資料型態再分為：整數、實數、浮點數等；但在 JS 裡，統統是 Number 資料型態。

### 3-9-1　宣告為 Number 物件

```
let n = 1.546;
```

## ▌ 3-10　常用屬性及內建函式

```
let n = 1.546;
toFixed(); // 四捨五入至小數點後指定位 回傳值為字串
n.toFixed(); // '2'
n.toFixed(1); // '1.5'
n.toFixed(2); // '1.55'
```

## ▌ 3-11　全域物件 MATH

Math 是個純物件，內建的屬性及函式，皆可以對應至常見的數學公式上，例如國中學過的圓周率 PI 以及三角函數 sin cos⋯的方法，皆可利用 Math 物件的內建函式來做呼叫使用。

## ▌ 3-12　常用屬性及內建函式

```
let n = 1.546;
Math.floor(n); // 1 無條件捨去至整數
Math.ceil(n); // 2 無條件進位至整數
```

```
let n = 9;
Math.sqrt(n); // 3 開根號 √
```

```
Math.PI // 回傳 pi 常數值 3.14....
```

```
Math.random(); // 隨機產生 0 至 1 之間的數
```

## 3-12-1 撰寫樂透 API

我們現在了解了 Number 和 Math，我們來試著替 node http server 新增一個出牌的功能

```
let getNo = (n)=> Math.ceil(Math.random()*n); // 從 1 至 n 隨機出一個數
```

完整程式碼：

```
var http = require("http");
const getNo = () => Math.ceil(Math.random() * 38); // 從 1 至 38 隨機出一個
數

//create a server object:
http
 .createServer(function (req, res) {
   res.write('' + getNo()); //write a response to the client
   res.end(); //end the response
 })
 .listen(8080); //the server object listens on port 8080
```

每當你重新整理 Browser 時

便會隨機產生 38 以下的值。

### 3-12-2 全域物件 Math 及 Number 的練習

輸入下面內容並印出值

```
console.log(Math.PI)
Math.round(4.7);
Math.round(4.4);

Math.pow(8, 2);
Math.sqrt(64);

Math.abs(-4.7);
Math.ceil(4.4);
Math.floor(4.7);

Math.sin(90 * Math.PI / 180);
Math.cos(0 * Math.PI / 180);

Math.min(0, 150, 30, 20, -8, -200);
Math.max(0, 150, 30, 20, -8, -200);
Math.random();
Math.abs(-7.25);
```

1.  若 25 是 (3x　2) 的正平方根，則 x= ?

2.  小康將一長木梯斜放在離牆 7 公尺處，此時梯頂與地面的距離為 24 公尺，求梯長多少公尺？

3.  7382　2622 = 2000 × a，則 a = ?

4.  試著將 樂透 API 引用 url , querystring 兩個內建模組，使其可以讓使用者指定最大值

## 3-13 全域物件 DATE & TIME

### 3-13-1 DATE 宣告方式

```
let today = new Date(); // 設定成今天的日期
```

## ＊ 宣告特定的日期

```
let d1 = new Date(2022,0,1); // 設定為 2022 年 1 月 1 日，注意月份是 zero-
base
let d2 = new Date(2022,0,1,16,27,45); // 設定為 2022 年 1 月 1 日 下午 4 點
27 分 45 秒
```

還有其他國家的約定俗成的日期的宣告方式，為避免混淆，先用公制。

```
let d3 = new Date(2022,1,0); // 但日期若設為 0，會指到前一天去，所以是 2022
年 1 月 31 日，這就要特別注意，月份是 zero-base 但其他的不是。
```

宣告從西元 1970 年 1 月 1 日 0 點 0 分 0 秒 至參數值的毫秒數，所以 1000
毫秒數 就是 1 秒

```
let d3 = new date(1000);
```

## ＊ 呼叫方式

```
today.getDate(); // 取得日期
today.getDay(); // 取得星期幾
… 不一一舉例
```

## ＊ 設定方式

```
d1.setDate();// 設定日期
… 不一一舉例
```

## ＊ 計算日期時間的區間

```
var date1 - new Date("06/30/2019");
var date2 = new Date("07/30/2019");

// To calculate the time difference of two dates
var Difference_In_Time = date2.getTime() - date1.getTime(); //getTime() 是取
出絕對時間

// To calculate the no. of days between two dates
var Difference_In_Days = Difference_In_Time / (1000 * 3600 * 24); // 是一天
的換算 1000 毫秒 * 3600 分鐘 * 24 小時
```

時間與日期大約是實務上最難搞的沒有之一，大家還記得 2000 年時，鼎鼎
大名的千禧蟲，當時的人們，擔心程式在判讀日期及時間上產生錯誤，進
而造成毀滅性的災難。

例如：航空系統錯誤，導致飛機相撞之類的，但幸好並未沒發生；但這也
告訴我們，遇到時間的處理，都要小心再小心。

例如你在台北買了一個傳統鬧鐘，設定早上 7 點響鈴叫你起床，一直運作
良好；但有一天你舉家搬到西雅圖，鬧鐘也帶去了，沒有做任何調整，然後，
你以為鬧鐘會一如往常在早上 7 點響鈴叫你起床，但因為時區的不同，這
並不會發生。

像這樣未確認時區的錯誤程式，在實務上，也是屢見不鮮。

所以混亂的來了

## ＊ 撰寫回報現在日期時間的 API

我們來試著替 node http server 新增一個出牌的功能

直接宣告一個 new Date(); 並且呼叫 toLocalString()

```
var http = require("http");

//create a server object:
http
 .createServer(function (req, res) {
   let today = new Date(); // 宣告今天的時間
   res.write(today.toLocaleString()); //印出時間
   res.end(); //end the response
 })
 .listen(8080); //the server object listens on port 8080
```

我在本地端 Chrome console 得到的是

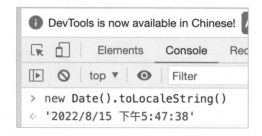

結果 browser 印出來的是：

很明顯是不相同的。

在以前的時代，因為軟體都是打包成應用程式在做販售，所以實務上，較不易發生這樣的問題，因為安裝上裝置之後，時間就是以裝置上的時間為準。

但進入雲端時代，所有解決方案幾乎都上雲，因此先確定好時區，變的格外重要。

例如台鐵有一個火車時刻表 API ，你要查當下的下一班列車的時刻，但因為你的雲端解決方案的時區不是台灣，所以你永遠給的是錯誤的當下時間，永遠查不到當下對的資訊，這樣類似的問題，在今時今日很常見，要特別注意。

那怎麼解決呢？

## ＊ 解決方法其一

加入參數 'zh-TW'

```
new Date().toLocaleString('zh-TW');
```

但現在需求改了，只要給我小時就好

```
new Date().getHours(); // 回傳的是當地的時間
```

而 toLocaleString('zh-TW') 的方法，沒有只回傳小時的，那該怎麼辦？

辦法其一，new Date().toLocaleString('zh-TW'); 拿到字串後，再用字串的 substring 方法去取出小時的位置；

```
let hours = new Date().toLocaleString('zh-TW').substring(11,13); // 拿到
小時的字串
```

但這不是個直覺的方法，可以想到日後可能會出現什麼其他的 bug，例如雲端解決方案的 node.js 版本不同，回傳的是繁體字的版本。

```
new Date().toLocaleString('zh-TW'); //'2022/5/22 下午 1:27:43'
```

那整個程式就會因為字串的位置錯誤，而取值錯了，取到的是英文字串或

是空字串,導致程式執行失敗而報錯,那還好辦,最怕的是取到其他的數字字串,導致程式還是可以運行,但是後面的資料是錯誤的,才會是大麻煩。

js 的開發是一個開放的領域,通常我會建議直接使用其他開發者寫好的套件,會是個好方法;下一節介紹套件 luxon

## 3-13-2 Date 練習

例如:

```
var d = new Date();
d.toLocaleString(); //=> "2020/12/3 下午 4:10:02"
```

輸入下列內容並印出值

```
var d = new Date(2018, 11, 24, 10, 33, 30, 0);
var d = new Date(2018, 11, 24, 10, 33, 30);
var d = new Date(2018, 11, 24, 10, 33);
var d = new Date(2018, 11, 24, 10);
var d = new Date(2018, 11, 24);
var d = new Date(2018, 11);
var d = new Date(2018); // 無效日期宣告 結果是 1970 年
var d = new Date(100000000000);
var d = new Date(-100000000000);
```

試著將 回報現在日期時間的 API 引用 url , querystring 兩個內建模組,使其可以讓使用者指定回到 2 天後

## * 引用 luxon

經過上一節的練習,是不是讓你開始感到混亂,其實有個簡單的方法,那就是不要自己寫,用套件,所以我們這裡引用的套件叫做 luxon

```
const {DateTime} = require('luxon');
```

利用其物件內件函式,就可以印出我們所期待的正確的時區的時間了

```
DateTime.now().setZone("Asia/Taipei").toString()
```

```
Browser

     ↻      https://shoorr.sse.codesandbox.io/

2022-05-19T18:21:05.607+08:00
```

如果我只要知道幾點

```
DateTime.now().setZone("Asia/Taipei").c.hour; // 只回傳 taipei 的幾點，c 有
全部的當地時間參數
```

## ＊ 計算時間日期

計算日期時間的區間等，是撰寫程式實務上，很常遇到的需求，當然可以用 Date 物件來做日期的計算等，但如上述理由，不建議，因為太容易出現 bug 了；所以用套件。

設定 2017 年 1 月 1 日

```
DateTime.local(2017, 1, 1); // 1 月就是 1，直覺多了
```

計算從 1982 年 5 月 25 日至今的 天數 及 小時

```
DateTime.now().diff(DateTime.local(1982, 5, 25), ['days', 'hours']) //[
Duration {"days":14604,"hours":19.727225555555556} ]
```

計算現在加上 15 分鐘 又 8 秒 是什麼時間

```
DateTime.now().plus({minutes: 15, seconds: 8}) //[ DateTime 2022-05-
19T19:58:46.010+08:00 ]
```

計算 2017 年 5 月 25 日跟 30 日 隔了幾天

```
DateTime.local(2017, 5, 30).diff(
    DateTime.local(2017, 5, 25),"days"
  ).days; // 5
```

除此之外，還有輸出格式的問題，例如 Date 物件的月份是 zero-base 的，這和人類的習慣也是不同的，難免也會有程式設計師在撰寫程式時，一時不查，導致邏輯上的錯誤，故這裡也是建議用套件。

例如我現在當下是 5 月

```
DateTime.now().c.month; // 5
new Date().getMonth(); // 4
```

像這樣的錯誤，就會因為一時的不查，而發生。

小結一下，程式撰寫時，因為 Date 操作錯誤，所導致的 bug 不勝枚舉，故建議不管日後寫的是前端還是還是後端，直接引用套件來做 Date 相關的撰寫，會是比較好的選擇。

## 3-14 陣列函式

和上述資料形態一樣，陣列的資料形態也有其內建函式，但不同的是，陣列的有些內建函式是會改變原本陣列的值。

例如 slice 和 splice 都是將陣列的值複製出來。

slice：

```
let ary = ['cat','dog','pig'];
let r = ary.slice(0 ,1 ); // 從第 0 個複製到 第 1 個 之前，都是 zero-base
r; // ['cat']
ary; // ['cat','dog','pig']; ary 原始值不變
```

splice：

```
let ary = ['cat','dog','pig'];
let r; = ary.slice(0 ,1 ); // 從第 0 個複製到 第 1 個 之前，都是 zero-base
r; // ['cat']
ary; // ['dog','pig']; ary 原始值的第一個不見了
```

在其他的資料型態的內建函式中，要改變原本變數的原始值，通常要像是如下的方式

```
let text = 'Hello World';
text = text.replace('World','Baby');
text;// 'Hello Baby'
```

但在陣列函式中，像這樣子呼叫之後，跟著改變原本變數的函式有很多，使用時要很注意，避免產生錯誤。

## 3-14-1 陣列宣告方式

```
let ary = ['cat','dog','pig'];
```

### ＊ 常用屬性及內建函式

```
// 取得陣列長度
ary.length; //3
// 取得該值第一次出現的位置
text.indexOf('dog'); //1 因為是 zero-base，所以是 1
// 陣列有無含該值
ary.includes('dog'); // true
// 輸出成字串
ary.join('_'); // 'cat_dog_pig' 用 _ 字符隔開，預設是 ，
```

### ＊ concat, push：

合併兩個或多個陣列

```
// concat vs push

let ary = ['cat'];
let r = ary.concat('dog');
r; // ['cat', 'dog'] ;
ary; //['cat'] 還是原來的值

let ary = ['cat'];
let r = ary.push('dog');
r; // 2 回傳陣列長度
ary; // ['cat', 'dog', 'pig'] 原始值被改變了
```

### 3-14-2 陣列練習

輸入下列內容並印出值

```
let ary = ['chicken','cat','dog','pig'];
var fruits = ['Strawberry','Apple', 'Banana'];
const array1 = ['a', 'b', 'c', 'd'];
const array2 = ['e', 'f'];
const array3 = array1.concat(array2);
const animals = ['pigs', 'goats', 'sheep','eagles'];

const count = animals.push('cows');

animals.push('chickens', 'cats', 'dogs');
```

## ▌ 3-15 自訂全域物件 & module

程式開發的習慣上，會把常用函數及常用的常數，寫成一個物件，並將該件物件做全域宣告。

看起來有像是 Math 物件，Math 有常數 PI ，也有常用的數學公式。

而這次我們寫的是自己的 Math 物件。

例如：

```javascript
let printValue = (v)=>{console.log(v)};

const Utils={
  printValue
}; // 常用的方法習慣上會被定義為 Utils

// 呼叫方式
Utils.printValue(123);
```

為了讓每個程式檔案都可以呼叫這個方法，我們會另外開一個新的檔案，寫好程式碼後，將其做為 module 輸出。

```javascript
module.exports = Utils;
```

開啟 codesandbox

1.  開一個新的 node-http-server 的 sandbox

2.  在檔案目錄區，點 📄 另開一新檔案 Utils.js

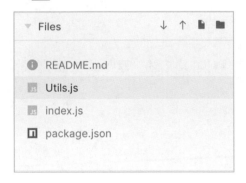

3.  將上述程式碼寫入 Utils.js 內

```
let printValue = (v) => {
  console.log(v);
};

const Utils = {
  printValue
};

module.exports = Utils;
```

4. 至 index.js 引用 Utils

5. 跟引用套件的方式一樣，只是這次是指向目錄內

```
const Utils = require('./Utils');
```

6. 就可以在 index.js 內呼叫 Utils 物件的所有屬性函式等。

```
Utils.printValue('hello');
```

## 3-15-1  exports 練習

試將前面章節的溫度轉換 及 樂透 等函式，撰寫成全域物件，並導入至 index.js 使用。

# 4

## JS 程式學習篇
## 決策與迴圈

## ▌ 4-1 學習目標／演練成果

學習條件判斷式及迴圈的處理

## ▌ 4-2 條件判斷式

條件判斷式簡單說，就是用程式碼的方式，來撰寫「選擇的流程」，例如：
判斷 60 分以上，叫做及格；不到 60 分就是不及格，那用程式要怎麼表達呢？

這邊可先利用 chrome 的 console 來做練習

```
if(score => 60 ){
  console.log(' 恭喜及格 !')
}else{
  console.log(' 不及格 !')
}
```

其中的 (score => 60 ) 就是用來評估的條件，如果符合條件，會回傳布林值
true 值，反之則是 false。

故，如果 score 是 60 以上則會真，即執行 if 所 { } 起來的程式區塊，反之，
則是會進入 else { } 程式區塊。

而在評估條件裡，通常可以置入算式運算子及邏輯運算子來協助我們做判
斷。

### 4-2-1 評估條件

\* 比較運算子

比較兩邊的數值

(70 => 60 )

流程

1. 70 大於 60 為真

2. 故回傳值為 ture

| 名稱 | 運算子 | 目的說明 | 範例 | 結果 |
|------|--------|----------|------|------|
| 等於 | == | 比較兩個數值是否相同 | 1 == 1 | true |
| 不等於 | != | 比較兩個數值是否不同 | 1 != 1 | false |
| 大於 | > | 左邊是否大於右邊 | 3 > 2 | true |
| 小於 | < | 左邊是否小於右邊 | 3 < 2 | false |
| 大於等於 | >= | 左邊是否大於或等於右邊 | 3 >= 2 | true |
| 小於等於 | <= | 左邊是否小於或等於右邊 | 3 <= 2 | false |
| 嚴格等於 | === | 比較兩個數值和資料型別是否相同 | 1 === 1 | true |
| 嚴格不等於 | !== | 比較兩個數值和資料型別是否不同 | 1 !== 1 | false |

等於跟不等於分兩種，原則上現在都是用嚴格等於跟嚴格不等於。

例如

```
1 == 1 ;// true
1 == '1' ;// true 值相同，資料型別不同， == 還是會回傳 true
1 === '1' ;// false  true 值相同，資料型別不同， === 會回傳 false
```

```
1 == true // true
'true' == true // false 以為 1 == '1' 會回 true ，結果這邊又回的是 false
```

因為用 == 與 != 比較容易出錯，所以現在都是用嚴格比較，如果真的需要
1 跟 '1' 比較的話，建議將 '1' 轉換成數字資料型態再做比較，例如：

```
1 === Number( '1') ; // true
```

就不會出錯了。

## ＊ 比較兩個運算式

```
(1+2)>(3-4); //true
```

也可以置入運算式做比較。

## ＊ 邏輯運算子

針對兩邊運算式的結果，再做邏輯判斷

```
((1<2)&&(5>4))
```

流程為：

1. 運算式 1(1<2) 為 true

2. 運算式 2(5>4) 為 true

3. 然後 AND 運算子再檢查兩邊均為 true

4. 故最後回傳值為 true

| 名稱 | 運算子 | 目的說明 | 範例 | 結果 |
|------|--------|----------|------|------|
| AND | && | 兩邊條件是否成立<br>true && true 為 true<br>true && false 為 false<br>false && true 為 false<br>false && false 為 false | (1<2)&&(5>4) | true |
| OR | \|\| | 兩邊條件一個成立<br>true \|\| true 為 true<br>true \|\| false 為 true<br>false \|\| true 為 true<br>false \|\| false 為 false | (1<2)&&(5<4) | true |
| NOT | ! | !true 為 false<br>!false 為 true | !(3 > 2) | false |

## 4-2-2 if 條件判斷式

```
if(score => 60 ){
  console.log(' 恭喜及格 !')
}else{
  console.log(' 不及格 !')
}
```

if 條件判斷式，判斷評估條件是否為 true，如果為 true ，則會執行 if 所 {
} 起來的程式區塊，反之則進入 else { } 的程式區塊。

除了 if… else 之外，還可以添加其他的 else if { } 條件，例如

```
if(score > 60 ){
  console.log(' 恭喜及格！')
}else if(score === 60 ){
  console.log(' 及格邊緣！恭喜 ')
}else{
  console.log(' 不及格！')
}
```

## ✳ if 條件判斷式簡寫

```
console.log((score > 60 ) ? ' 恭喜及格！':' 不及格！')
```

將 ? 當成 if，前面的 (score > 60 ) 依舊是評估條件，回傳值則是用 : 隔開，前面是評估條件為真 true 時執行，後面則為假 false 時執行。

## ✳ if 條件判斷式練習

```
let hour = new Date().getHours();
if ( hour < 18) {
  greeting = "日安";
} else {
  greeting = "晚安";
}
console.log(greeting);
```

```
let hour = new Date().getHours();
if ( hour < 9) {
  greeting = "早安";
} else if ( hour < 17)  {
  greeting = "午安";
} else {
  greeting = "晚安";
}
console.log(greeting);
```

```
let hour = new Date().getHours();
if ( hour < 17) {
  greeting = "午安";
}else if( hour < 9) {
  greeting = "早安";
} else {
  greeting = "晚安";
}
console.log(greeting);
```

## 4-2-3 switch 條件判斷式

```
switch(level){ // 依照 level 的值，來做對應
  case 1: // level 是 1 的話，才會進這個區塊
    console.log('好棒棒！')
    break; // break 有加才會退出 switch 判斷式，否則會執行下去
  case 2:
    console.log('優秀！')
    break;
  case 3:
    console.log('不錯喔！')
    break;
  default:
    console.log('別氣餒！')
}
```

switch 條件判斷式，依照 switch 所指定的變數來做相關的判斷，以變數值來做程式區塊的對應處理，加上 break 來斷掉程式區塊的執行。

所以實務上，很常利用 break 的特性，來讓程式簡潔易讀；例如，有些條件成立時，會執行一樣的程式碼；就會透過故意在 case 後，不加 break 來讓程式可以不被中斷。

例如：

```
let rep;
switch(text){
  case 'hi':
  case 'hello':
  case ' 你好 ':
    rep= ' 你好，最近過的如何？ '; // 因為回應都是相同的，故前面故意不加 break
    break;
  default:
    rep = ' 沈默以對 ...'
}
```

switch 條件判斷式練習

```
switch (new Date().getDay()) {
  case 0:
    day = " 星期日 ";
    break;
  case 1:
    day = " 星期一 ";
    break;
  case 2:
    day = " 星期二 ";
    break;
  case 3:
    day = " 星期三 ";
    break;
  case 4:
    day = " 星期四 ";
    break;
  case 5:
    day = " 星期五 ";
    break;
  case 6:
    day = " 星期六 ";
}
```

## 4-2-4 捷徑 / 簡寫

JS 目前有很多這類的語法。

這類的語法都是後來為了撰寫上的需求所發展出來的，例如先有函式，後來又有箭頭函式，箭頭函式可以增加程式的易讀性。

但對於初學者來說，一開始就這樣寫，會很容易寫錯，建議先看得懂就好，等未來對於 JS 更加熟悉時，再做這樣子的簡寫。

例如：

```
let rep;
let hello = 'hello';
if(hello){
  rep = hello;
}else{
  rep ='unkown';
}
```

hello 如果有值的話， rep 就 = hello ，沒有的話，就是 'unkown' 。

可以簡寫成

```
let hello = 'hello';
let rep = ( hello || 'unkown');
```

因為 || 是邏輯運算子，會先判斷 左邊是否有值，有的話，就會先回傳左邊，沒有的話，則會回傳右邊；所以就可以兩行寫完上面的程式碼。

## 4-2-5 直值與假值

例如

```
let n = (0 || 1); // 1
```

在上述的例子裡，你可能會覺得 0 是個數值，怎麼會回傳 1 ，那是因為 JS 有定義一些值為假值 false 。

| 假值 | 說明 |
|---|---|
| `let n = false;` | 布林值 false |
| `let n = 0;` | 數字 0 |

| let n = ''; | 空白值 |
| --- | --- |
| let n = 10/'s'; | NaN（不是一個數字） |
| let n ; | 未定義 undefined |

上述假值階會回傳 false，而除了上述以外，其他幾乎皆為真值 true。

## 4-2-6　捷徑及真假值練習

```
let name1 = ' 皮耶 ';
name1 += ' 你好 ';
```

```
const length = data && data.length
```

```
var a = 10, b = 0, c = ' 前端 ';
a || b && (c = ' 後端 ');
console.log(c);
```

```
var a = 5, b = 0, c = ' 前端 ';
(a || b) && (c = ' 後端 ');
console.log(c);
```

```
var a = 40, b = 30;
c = (a < b) && (a = 15);
console.log(a,c);
```

```
var a = 20, b = 30;
c = (a < b) || (a = 15);
console.log(a,c);
```

```
var a = 40, b = 30;
c = (a < b) || (a = 15);
console.log(a,c);
```

# ▎4-3　迴圈

迴圈設計出來的目的就是為了盡量減少重複的呼叫。

例如需要印出 1 到 5：

```
console.log(1);
console.log(2);
console.log(3);
console.log(4);
console.log(5);
```

但藉由迴圈，可以寫作

```
for(let i = 1; i<= 5 ; i++){
  console.log(i)
}
```

依舊可以印出 1 到 5，但程式碼變的簡單也明確許多。

初學者要小心的是，因為滿足條件的邏輯錯誤，造成無窮迴圈而讓程式崩潰喔。

例如 (let i = 1; i > -1 ; i++) 因為 i 永遠大於 -1，故會一直執行程式區塊內的 console.log(i) 而造成程式崩潰。

迴圈在 JS 的語法裡，主要由 for , while , do while 來實現；但當你對 JS 語法的熟悉度越來越高時，其實比較常用的會是陣列迴圈。

## 4-3-1 for 迴圈

```
for(let i = 1; i <= 5 ; i++){
  console.log(i);
}
```

條件及計數器為 (let i = 1; i<= 5 ; i++) 由 ; 隔開。

第一項 let i = 1; 是宣告初始值的地方，可以去掉 let 關鍵字。但筆者建議有宣告的關鍵字，會比較好閱讀。

而這裡也沒有規定，只能在這裡宣告一個變數，也可以宣告多個變數，例如：let i = 1,j =  2 , s='hello'; 等第二項 i <= 5; 為跳出條件，這裡的意思是，當 i 大於 5 時，就會跳出迴圈。

同樣，這裡也沒有規定只能放一個條件，也是可以放 i <= 2,i < 10 之類的，但是想要跳出迴圈，條件就必須同時滿足這兩個，假設你設了 i <= 2,i > 5 之類的，因為永遠不可能滿足，那就會變成無窮迴圈了。

第三項 i++ 則為更新的項目，當每次執行完 { } 程式區塊後，便會執行 i++，然後當 i = 6 時，變會跳出這個迴圈了；一樣也是可以放兩個以上的更新項目。

例如：

印出 9 9 乘法表

```
for (let i = 1; i < 10 ; i++){
  for (let j = 1; j < 10 ; j++){
    console.log(i,j,i*j);
  }
}
```

可以被改寫成

```
for(let i =1, j =1 ; i < 10 ; (j==9) ? i++:null,(j==9) ? j=1:null,j++){
    console.log(i,j,i*j);
}
```

雖然行數變短了，但程式閱讀上變得更不易了。

我們使程式碼變短的目地，通常是為了增加易讀性；但像這個，這樣子做，反而變的更難閱讀了，雖然看起來很炫砲，但筆者就不建議了。

## 4-3-2 while 迴圈

```
let i = 0;
while(i<5){
  console.log(i);
  i++;
};
```

(i<5) 是條件,當 i 大於 5 時,就會跳出迴圈;尚未滿足條件時,會執行 { } 程式區塊。

### 4-3-3　do while 迴圈

```
let i = 0;
do {
  console.log(i);
  i++;
}while(i<5)
```

先執行 { } 程式區塊,然後遇到 (i<5) 為條件,當 i 大於 5 時,就會跳出迴圈。

### ＊ break

```
let i = 0;
while(true){
  console.log(i);
  i++;
  if(i>=5){break;}; // 當 i >= 5 時,變會跳出迴圈
};
```

### ＊ continue

```
let i = 0;
while(i<5){
  i++;
  if(i===3){continue;}; // i 等於 3 時會跳過執行下面的程式,所以不會印出 3
  console.log(i);
};
```

### 4-3-4 迴圈練習

```javascript
for (let i = 0; i <= 10; i++) {
 if (i === 3) {
   continue;
 }
 console.log(i);
}
```

```javascript
var i = 0;
while (let i <= 10) {
  if (i === 3) {
    continue;
  }
  console.log(i);
  i += 1;
}
```

## ▎ 4-4 陣列函式的迴圈應用

迴圈的最常處理的就是資料的計算，例如我們有一群學生的分數資料，想要替他們加上及格不及格的字樣

```javascript
let students = [
{
  name:'amy',
  score:79
},{
  name:'peter',
  score:25
},{
  name:'john',
  score:90
}
]
```

如果是傳統迴圈的撰寫方式

```
for (let i = 0 ; i < students.length; i++){
    if(students[i].score >= 60){ // 分數比 60 分大
        students[i].comment = ' 及格 ';
    }else{
        students[i].comment = ' 不及格 ';
    }
}
```

但藉由陣列內建的迴圈函式 map , filter , reduce 等，可以讓我們的程式更簡潔易讀。

## 4-4-1 map

map 函式的參數為函式，map 會以陣列的長度 ( length) ，去呼叫函式的次數，每次會依序倒入陣列的值，如下

```
students.map(
  (v,i,ary)=>{
// v 是依序陣列的值，第 1 次呼叫就是 {name: 'amy', score: 79} 第 2 次就是
{name: 'peter', score: 25}
// i 是第幾個 zero-base ，故第 1 次的 i 是  0
// ary 是原始值，也就是陣列 students
    if(v.score >= 60){
        v.comment = ' 及格 ';
    }else{
        v.comment = ' 不及格 ';
    } // 分數比 60 分大
  }
)
```

比用 for 迴圈撰寫簡潔多了

map 也可以直接回傳陣列出去，例如，我們不想知道分數，只想知道每個人有沒有及格

```javascript
let comment = students.map((v)=>{
    if(v.score >= 60){
     return '${v.name} 及格 ';// return 值回去
    }else{
       return '${v.name} 不及格 ';// return 值回去
    }
})
comment; //['amy 及格 ', 'peter 不及格 ', 'john 及格 ']
```

要注意的是 map 的回傳機制是就算沒有 return 值出去，它也會 return 一個 undefined 出去，假設不 return 不及格：

```javascript
let comment = students.map((v)=>{
    if(v.score >= 60){
     return '${v.name} 及格 ';// return 值回去
    }
})
comment; //['amy 及格 ', undefined , 'john 及格 ']
```

comment 變成 ['amy 及格 ', undefined, 'john 及格 ']

我們不想要 undefined 怎麼做呢，可以試試 filter

## 4-4-2 filter

現在我們只要及格名單

```javascript
let pass = students.filter(
    (v,i,ary)=>{ // 跟 map 的參數是一樣的設定
        if(v.score >= 60){
            return true;// return 任何真值 , 都是 return v 回去 , 這裡跟 map 不
一樣
        }
    }
)
```

就可以拿到及格名單了，但我們只想要名字，不想要其他的資料，可以試試 reduce

### 4-4-3 reduce

```
let pass = students.reduce(
  (a,b,i)=>{
    // i 是執行第幾圈了，所以一開始是 1
    // a 在第 1 圈時，是第 [0] 值，之後是上一圈的回傳值
    // b 在第 1 圈時，是第 [1] 值，之後是下一個值
    let r = [];
    if(typeof(a)==='object' && typeof(a.name)==='string'){ // 第 1 圈會是
{name: 'amy', score: 79}    所以要判斷一下型別
        if(a.score >= 60){
            r.push(a.name)
        }
    }else{ // 第 2 圈之後變成上一圈回傳值
        r = a; // 直接等於上一圈的回傳值
    }
    if(b.score >= 60){
        r.push(b.name)
    }
    return r;
  }
)
pass ; // ['amy', 'john']
```

reduce 在邏輯上比較複雜，但如果用的好的話，可以讓程式更簡潔。

例如 (typeof(a)==='object' && typeof(a.name)==='string') 這一行，可以再改成 (a && a.name) 又更簡潔了。

### 4-4-4 sort 排序

原則上是泡泡排序 (Bubble Sort) 由左而右、兩兩比較相鄰資料，若前者大於後者，則將兩者交換來完成排序。

```
let arj - [ 2, 3, 4, 1];
let r = arj.sort((a, b)=>{
//a 是後值
//b 是前值
 return a - b;
//return -1 以上會交換位置
// 第一次 3-2=1 不動
// 第二次 4-3=1 不動
// 第三次 1-4=-3 1往前比
// 第四次和前面的 3 比 1-3=-2 1再往前排
// 第五次和前面的 2 比 1-2=-1 1再往前排
// 前面沒了
});
r; // r 是 [1, 2, 3, 4]
a; // a 也是 [1, 2, 3, 4]   原始值被改變了
```

但是以前述 students 為例，我想要依照學生的成績排序，但我不希望原始 students 的陣列值被改變，那就得將 students 複製出來，但直接複製陣列的值，在 JS 裡對應的是陣列的參考位置 reference 。

```
let ary = [ 2, 3, 4, 1];
let n = ary;
ary.sort((a, b)=>a-b);
```

ary 變成 [1, 2, 3, 4] 沒問題，但你會發現 n 也變成了 [1, 2, 3, 4] 。

這是因為 JS 為 複合資料型態時（ Object / Array / Function），運算式其實都是傳址（pass by reference）運算，這也可以解釋，為什麼前述陣列函式，例如 concat 就不是傳址運算，所以不會去改動到原本的變數。

如果要避免這個問題，可以用 JSON.parse(JSON.stringify(ary)) ，也就是先將 ary 轉成字串後，再將字串下 JSON.parse 去轉成 Array 。

## 4-4-5 陣列函式的迴圈應用練習

```
[5, 6, 7].map( v => v + 1 )

let array = [2, 3, 6, 8] ;
let newArray = array.filter(item => item !== 6) ;
console.log(newArray);

const arr = [2, 4, 6];
let total = 0;
for (let i = 0; i < arr.length; i++) { total += arr[i]; }; console.
log(total);

const arr = [3, 6, 9];
const sum = arr.reduce((acc, current) => acc + current,0);
console.log(sum);

const arr = [1, 2, 3];
const mapArray = arr.map(value => value * 2)
```

試排序下列陣列變數

```
var ary1 = [148,768,434]; //=> [148,434,768]
var ary2 = ["a","acdsas","efasd"];
var ary3 = ["148",768,434];
var ary4 = [148,768,434,433,78977,4345,435,47837];
```

# 5

## JS 程式學習篇
## 非同步與 API

## ▌ 5-1 學習目標／演練成果

學習使用非同步語法

1.　Promise 物件
2.　await & async

## ▌ 5-2 同步 sync vs 非同步 async

### 5-2-1 何謂同步？

到目前為止學到的程式，通通都是同步，所謂同步，就是一行執行完成後，才會跟著執行下一行。

例如：

```
console.log(1);
console.log(2);
console.log(3);
// 一步一步執行，依序印出 1 2 3
```

或是

```
let a = 1;
let b = 2;
let ans = a + b;
```

電腦會先宣告 a 再宣告 b 然後加總得到 ans。

就算將加總先宣告成函式，再呼叫，例如：

```
let getAns = (...items)=> items.reduce((a,b)=>a+b)
let a = 1;
let b = 2;
let ans = getAns(a,b);
```

雖然看似函式在一開始就被宣告，在最後才被呼叫，但這些程式碼，依然逐行執行的，也就是同步執行。

## 5-2-2 何謂非同步？

在最早的 JS 語法中，要實現非同步，會藉由 setTimeout 來實作，例如：

```
console.log(1);
setTimeout(()=>console.log(2),1000); // 在 1 秒過後執行
console.log(3);
// 依序印出 1 3 2
```

雖然 setTimeout(... ,1000); 是寫在第二行，也是第二個被執行的，但因為設定為延遲 1 秒，才會去執行函式，故它會等個 1 秒，才會去執行程式 console.log(2)，所以 setTimeout 是 JS 非同步語法，到這裡還很容易理解。

但是，我如果將 1 秒，改成 0 秒呢？

```
console.log(1);
setTimeout(()=>console.log(2),0); // 在 0 秒過後，執行 ()=>console.log(2)
console.log(3);
// 依序印出 1 3 2
```

結果依然是印出 1 3 2，而不是 1 2 3。

而這又是為什麼呢？

## 5-2-3 阻塞（blocking）

這就得談為什麼要有「非同步」的程式設計，前述的「同步」就是一步一步的運行程式，一行執行完畢，才會跟著執行下一行：

例如：

```
console.log(' 工作 1 完成 ');
for(let i = 0 ; i<=5000000 ; i++){
  if(i===5000000){console.log(' 工作 2 完成 ');}
}
console.log(' 工作 3 完成 ');
```

很明顯程式的執行必須等待工作 2 完成，才能執行 工作 3 ，這樣子等待，使得主線程 main thread 整個等 工作 2 執行完，才能執行 工作 3 ，就是所謂的阻塞（blocking）。

但其實工作 2 有無執行完成，並不妨礙工作 3 的進行，所以為了解決阻塞（blocking）的問題，非同步語法 setTimeout 登場，可以先將 工作 2 移到 Timer 去，不去防礙主線程 main thread 的運行。

```javascript
console.log(' 工作 1 完成 ');
setTimeout(()=>{
  for(let i = 0 ; i<=5000000 ; i++){
    if(i===5000000){console.log(' 工作 2 完成 ');}
  }
},0);
console.log(' 工作 3 完成 ');
```

## 5-2-4 單線程 single threaded

其實 JS 從頭到尾，只有一個單線程可以作運算，為了模擬出非同步，也就是使用 setTimeout 這類非同步語法，使其看似雙線程以上的運算，故設計出了 Call Stack & Task Queue 的架構。

## ＊ 什麼是 Call Stack ？

例如：

```javascript
function f1(){console.log(1);}
function f2(){f1();}
function f3(){f2();}
f3();
```

當呼叫 f3 函式時，會將 f3 函式丟進 Call Stack，然後再丟入 f2 進 Call Stack，再丟入 f1 進進 Call Stack。

|    |    | f1 |
|----|----|----|
|    | f2 | f2 |
| f3 | f3 | f3 |

當函式都進入 Call Stack 後，才會開始執行。

會先執行最後進來的 f1 函式，f1 執行完畢後，再執行 f2 ，f2 執行完畢，再執行 f1，f1 執行完畢，結束。

| f1(); | | |
|---|---|---|
| f2 | f2(); | |
| f3 | f3 | f3(); |

這就是所謂 Stack ，就是先進後出的架構，所以最後進來的 f1 ，最先被執行。

## 5-2-5 堆疊追蹤

附帶一提，當我們了解 Call Stack 的概念後，當程式出現 bug 時，就可以跟據 error 訊息去追蹤錯誤。

例如：

```
function f1(){console.log(a);} // 沒有宣告 a 變數
function f2(){f1();}
function f3(){f2();}
f3();
```

當程式執行時，變會中斷，並產生錯誤訊息：

```
ReferenceError: a is not defined // 未宣告 a 變數
    at f1 (/sandbox/index.js:1:15) // 在 f1 函式（第 3 行第 15 個字元）
    at f2 (/sandbox/index.js:2:3)// 在 f2 函式（第 6 行第 3 個字元）
    at f3 (/sandbox/index.js:3:3)// 在 f3 函式（第 9 行第 3 個字元）
```

當錯誤發生時，錯誤訊息告訴我們，因為未宣告 a 變數，所以程式中斷了，在 f1 函式裡，而 f1 是被 f2 所呼叫執行，所以下一行也告訴我們，f1 的上一層是被 f2 函式呼叫，那 f2 也是被 f3 所呼叫。

這裡就可以很明白的看出何謂 Call Stack 。

＊ 錯誤處理

附帶一提，程式在運行時，難免會有一些狀況，是我們沒有辦法去掌握，而導致程式錯誤中斷。

但有的時候，我們本來就知道會有錯誤發生的機率，但我們並不希望程式因此而中斷，我們就可以設計一些方式，來避免程式中斷，這個叫做程式防呆。

例如，上述的程式會中斷報錯

```
try {
  f3();
}catch(err){
  console.log()
}
```

但只要加上 try{ } catch(){ } 程式便不會中斷下來，這個技巧在預期不穩定的地方很實用，例如 fetch api 等。

## 5-2-6 非同步語法 (setTimeout)

```
console.log(' 工作 1 完成 ');
setTimeout(()=>console.log(' 工作 2 完成 '),0);
console.log(' 工作 3 完成 ');
```

而 setTimeout 非同步語法，在前端環境用的是 Web API ，在後端環境 node.js 用的是 Timer 模組；先將工作 2，丟到 Web API / Timer 那裡，等到設定的時間到了之後，將工作 2 丟到 Task Queue 裡，等待 JS 的呼叫；

＊ **Task Queue 工作序列**

當 Call Stack 是空的，沒有工作需要被執行時，這時候 event loop 就會去 Task Queue 裡，找找有沒有工作需要被執行，有的話，就會將工作拉到 Call Stack 裡去執行。

## * event loop 事件循環

loop 就是迴圈，event loop 可以看做是一個不斷檢查 Task Queue 有沒有工作需要被丟到 Call Stack 的無限迴圈。

## * 整體流程

```
console.log(' 工作 1 完成 ');
setTimeout(()=>console.log(' 工作 2 完成 '),0); // 先被丟到 Web API / Timer
，再立刻被丟到 Task Queue
console.log(' 工作 3 完成 ');
// 依序印出 工作 1 完成 工作 3 完成
// 主線程工作執行完畢， event loop 啟動，找到在 Task Queue 的工作 2 ，立刻拉回
Call Stack
// 印出 工作 2 完成
```

理解了這整體是怎麼運作之後，再回來非同步語法的程式撰寫的部份。

在寫程式時，雖然會有非同步語法的需求，但在實際撰寫時，用 setTimeout 這樣子的語法，在程式維護上是不理想的，因為不夠直覺。

尤其是 AJAX( Asynchronous JavaScript and XML) 問世之後，撰寫非同步程式的機會變得更多，也更加複雜起來。

因此為了使我們撰寫非同步的程式時，可以增加程式的維護性及易讀性，Promise 物件及 await & async 的語法就問世了。

前述的程式，可以改成：

```
let p = new Promise((resolve,reject)=>{setTimeout(()=>resolve(' 工作 2 完
成 '),0);});

(async()=>{
  console.log(' 工作 1 完成 ');
  console.log(await p);
  console.log(' 工作 3 完成 ');
})();
```

執行之後，會依序印出 1 2 3。

程式也變的明瞭清楚許多。

### 5-2-7 AJAX( Asynchronous JavaScript and XML)

AJAX 會發展起來，是因為以前頻寬是很珍貴的，那時的人覺得在表單填完送出資料之後，伺服器再回傳回來的網頁，大部份跟之前表單的網頁，大部份都是一樣的網頁結構，那何不節省點，只和伺服器做資料的交換就好呢？ 例如 google 關鍵字時

當我們輸入關鍵字時，在網頁的背景中，已經在執行 ajax 的程式來跟伺服器做溝通，當獲得 google 所建議的資料時，就會直接印在原網頁上，而不需要再重新整理網頁，來更新資訊，這個技術，就叫做 AJAX 應用。

## ▎5-3 非同步語法的實現

因為 AJAX 的流行，所以非同步語法成為主流。

### 5-3-1 XMLHttpRequest

```
const xhr = new XMLHttpRequest(); // 宣告 XMLHttpRequest 物件
    xhr.onreadystatechange = function () { // 當 xhr 進入
onreadystatechange 狀態時, 所要處理的事情。
        if (xhr.readyState == 4 && xhr.status == 200) {
           console.log(xhr.responseText);// 印出網址的內容
        }
    };
    xhr.open("GET", "api", true);// 我要訪問 api 這個網址
    xhr.send();// 執行
```

這是最早為了實現 AJAX 的語法,可以看的出來相當的不直覺,程式邏輯要處理網址內容的部份,寫在中間,裡面還是巢狀結構,在維護上並不直覺,所以後來 fetch 函式問世。

### 5-3-2 Fetch

```
fetch('api')
  .then((r) => {
    return r.json();
  })
  .then((r) => {
    console.log(d);
  });
```

因應著 XMLHttpRequest 的問題,fetch 函式問世了,如前述,同樣的需求, 使用 fetch 函式訪問網頁時,會回傳的是 Promise 物件, 看起來更簡潔了,但在維護上,仍舊不夠直覺,因為成了一圈一圈的巢狀結構了,能不能讓其逐行執行呢? 這樣會更好維護;然後搭配 Promise 的關鍵字 async & await 誕生了。

### 5-3-3 Async & Await

```
(async () => {
  let p = fetch('api');
  let r = await p.then();
  let d = await r.json();
  console.log(d);
})();
```

promise 物件搭配 async & await 關鍵字，實現了逐行結構，更加深易讀與維護性了。

XMLHttpRequest 所使用的 callback 的程式寫法，以及 fetch 剛問世時，所流行的 CPS 的程式寫法，本書不再作探討，因為已經過時了。

本書將就 promise 物件搭配 async & await 關鍵字的部份來做學習與練習。

## 5-4 Promise 物件與 async & await 語法

前面解釋了這麼多，終於正題，非同步相關的物件及語法，要如何做宣告及使用呢？

### 5-4-1 Promise 物件宣告方式

```
let p = new Promise((resolve,reject)=>{ })
```

Promise 物件是一個建構函式，產生一個 Promise 物件，參數必須為函式，而這個函式必須依照 js 的規定的格式 (resolve,reject)=>{}，才能正確的使用。

resolve 代 表 的 是 執 行 成 功，可 以 帶 任 何 的 回 傳 值，例 如 resolve({word:'hello'})

reject 代表的是執行失敗，可以宣告 Error 物件回傳 例如 reject(new Error('error'))

## ＊ Promise 的狀態

promise 有 3 種狀態：

1.   pending 擱置：初始狀態。

2.   fulfilled 實現：表示操作成功地完成。

3.   rejected 拒絕：表示操作失敗了。

我們以一個簡單的倒數計時器為例

```
let getTimer =(isOk)=> new Promise((resolve,reject)=>{
  setTimeout(()=>{isOk ? resolve('done'):reject(new Error('error'))},3000)
});
let timer1 =  getTimer(true);
timer1 // 當下是 Promise {<pending>}
timer1 // 3 秒後是 Promise {<fulfilled>: 'done'}
let timer2 =  getTimer(false);
timer2 // 當下是 Promise {<pending>}
timer2 // 3 秒後是 Promise {<rejected>: Error: error}
```

Promise 物件的狀態會隨著 resolve 及 reject 的被呼叫而轉變，最開始的狀態是 pending ，當 resolve 被呼叫後，狀態就會變成 fulfilled ，當 reject 被呼叫後，狀態會變成 rejected。

## ＊ Promise thenable

我們以 getTimer 函式產生的 timer1 Promise 物件為例。

```
getTimer().then(
  (v)=>{}, // resolve
  (err)=>{} // reject
)
```

Promise 物件，後面可以呼叫 then 函式，可以塞入兩個函式參數，第一個為 resolve 發生時，也就是預期的工作完成時，就會進入第一個函式；而第二個則是 reject 發生時，會被呼叫。

但目前會被改寫為：

```
getTimer(true).then( // 完成的話，進來這裡
  (v)=>{doSomeThing(v);},
).catch( // 錯誤的話，進來這裡
  (err)=>{}
)
```

程式碼會變得較易讀。

## ✳ **Promise Chain 鏈接**

```
timer1.then(
  (v)=>{return doSomeThing1(v)},
).then(
  (v)=>{return doSomeThing2(v)},
).then(
  (v)=>{return doSomeThing3(v)},
)
```

Promise 的 then 特性，可以直接從面一直加 then 下去，但是記得要回傳參數，並且為 Promise 資料型態，後面才能得到回傳的參數。

一樣這個 CPS 的程式寫法已經較退流行了，不過在很多的 framework 還是會看到，必須了解一下。

但 then 的巢狀結構，在程式維護上，其實不太理想，改成逐行比較易讀。

```
(async()=>{
  let r1 = await timer1().then();
  doSomeThing1(r1);
  let r2 = await timer1().then();
  doSomeThing1(r2);
  let r3 = await timer1().then();
  doSomeThing1(r3);
})()
```

## 5-4-2 await & async 宣告方式

```
function f0(){};
async function f1(){};
let f2 = async ()=>{}; // 箭頭函式
```

async 函式被呼叫時，會回傳一個 Promise 物件

```
f0(); //undefined
f1(); //Promise {<fulfilled>: undefined}
f2(); //Promise {<fulfilled>: undefined}
```

### * await

await 運算子被用來等待 Promise，只能在 async function 內使用。

```
await f1();
await f2();
```

實務上 await 常會拿來等非同步 API 的呼叫。

例如

```
let response = await fetch(url);
```

### 5-4-3  Async & Await & Promise 練習

```
function resolveAfter2Seconds(x) {
  return new Promise(resolve => {
    setTimeout(() => {
      resolve(x);
    }, 2000);
  });
}


async function add1(x) {
  const a = await resolveAfter2Seconds(20);
  const b = await resolveAfter2Seconds(30);
  return x + a + b;
}

add1(10).then(v => {
  console.log(v);   // prints 60 after 4 seconds.
});


async function add2(x) {
  const p_a = resolveAfter2Seconds(20);
  const p_b = resolveAfter2Seconds(30);
  return x + await p_a + await p_b;
}

add2(10).then(v => {
  console.log(v);   // prints 60 after 2 seconds.
});
```

## ▌ 5-5  並行運算 concurrent computing vs 平行運算 parallel computing

前面講述非同步語法時，講述了 JS 設計了 Call Stack & Task Queue & event loop 的機制，來模擬出多工環境，但其實只有單執行緒在做運算，這個設計就是所謂的並行運算 concurrent computing。

會有這樣設計的原因,就是因為 JS 本來一開始就是在前端瀏覽器的語法,因為瀏覽器環境的限制,當然只會有單執行緒,但是到了後端 node.js 開發時,其實可以用的資源就變多了,包括執行緒的資源都是可以被叫用的。

這使得平行運算 parallel computing 多執行緒運算,在後端 node.js 的環境中,是可以被實現的。

實作 node.js 的多執行緒運算:

1. 開一個新的 node-http-server 的 sandbox

2. 新增一個 main.js 及 worker.js

main.js:

```javascript
const { Worker } = require("worker_threads"); // 引用 worker_threads

const runService = (workerData) => {
 return new Promise((resolve, reject) => { // 建立 Promise 物件
   const worker = new Worker("./worker.js", { workerData }); // 讀取檔案
worker.js 的程式碼,並且倒入 workerData 參數,並且新增執行緒運行之。
   worker.on("message", resolve); // 當 worker 處理完後,會倒 message 過來,
這個時候呼叫 resolve 完成 Promise。
 });
};
(async () => {
 const result = await Promise.all([ //all 函式會等全部的 Promise 完成
   runService({ // 呼叫 runService 函式,並倒入參數
     a: 1,
     b: 2
   }),
   runService({
     a: 3,
     b: 4
   }),
   runService({
     a: 5,
     b: 6
   })
 ]);
 const sum = result.reduce((a, b) => a + b); // 將結果加總
 console.log(sum);// 印出結果
})();
```

worker.js:

```
const { workerData, parentPort } = require("worker_threads");
// workerData 為母進程傳送過來的資料
const { a, b } = workerData;
let sum = a + b; // 相加
console.log('worker.js ${a}+${b}=${sum}');
parentPort.postMessage(sum); // 回傳運算結果回母進程
```

利用 worker_threads 將 JS 的檔案，讀進來後，運行至新的線程上，這就是平行運算 parallel computing。

# 6

實作練習篇
實作網站版剪刀
石頭布

## ▌ 6-1 學習目標／演練成果

本章將會藉由實作一個簡單的剪刀石頭布的網站來了解，目前所學到的程式技術可以作到怎樣的應用，並在此實作中釐清前後端觀念。

第一階段的目標是實作從後端運行的剪刀石頭布。

第二階段的目標是則是將其邏輯搬到前端，讀者可以從實作中理解何謂前端及後端。

第三階段，使用後端套件 express 前端套件 webpack & react 來開發，理解使用套件的好處。

## ▌ 6-2 使用者輸入參數

1.  至 codesandbox 開 node http server ，安裝完 nodemon 且改寫 package. json 裡的 start 為 nodemon index.js

2.  引用 url 及 querystring 讓我們可以獲得使用者所輸入的參數。

```
const url = require('url'); // 引用內建模組 url
const querystring = require('querystring'); // 引用內建模組 querystring
```

3.  res.write 前加上 res.writeHead(200, { "Content-Type": "text/html; charset=utf-8" });

4.  這個是告訴來訪問的瀏覽器，這個伺服器將用 utf-8 輸出。

5.  另外在 function (req, res) {..} 裡，加上

```
let { query } = url.parse(req.url);
let { user } = querystring.parse(query);
```

拿到 user 所輸入的資料

6.  後面再加上

```
  if (user) {
    res.write(user); //write a response to the client
  } else {
    res.write(' 無資料 ');
  }
```

當使用者有輸入 user 的資料時，就印出來，沒有就印 ' 無資料 '。

因此 index.js 目前為：

```
var http = require("http");
const url = require('url'); // 引用內建模組 url
const querystring = require('querystring'); // 引用內建模組 querystring

//create a server object:
http
 .createServer(function (req, res) {
   let { query } = url.parse(req.url);
   let { user } = querystring.parse(query);

   res.writeHead(200, { "Content-Type": "text/html; charset=utf-8" });
   if (user) {
     res.write(user); //write a response to the client
   } else {
     res.write(' 無資料 ');
   }
   res.end(); //end the response
 })
 .listen(8080); //the server object listens on port 8080
```

7.   可以試著在網址列後輸入  /?user= 石頭

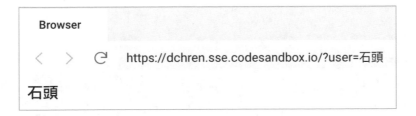

就會得到使用者所輸入的值，如果沒有輸入值，就會印出無資料。

## 6-3 新增 GAME 物件

1. 至 Files 下新增 game.js 檔案

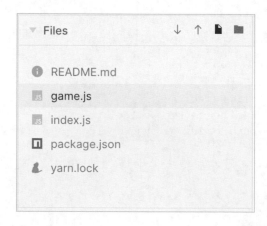

將所有跟遊戲相關的邏輯都寫在 game.js 裡，避免 index.js 過於繁雜，盡量讓一個程式的檔案只負責一件事情，例如 index.js 只負責伺服器的部份，而 game.js 負責遊戲邏輯的部份，在維護上會比較方便。

2. 先 game.js 宣告 GAME 物件，並且 export 出去

```javascript
const GAME = {
 play: (user) => { // 遊戲開始
   if(!user){ // 如果沒有值的話，直接 return 掉
     return;
   }
   console.log(' 使用者輸入 ${user}'); // 印出使用者輸入的資料
 }
};
module.exports = GAME;
```

3.   在 index.js 裡 引用 game

```
const GAME = require('./game');
```

4.   並且呼叫

```
let result = GAME.play(user);
```

5.   並且預期 play 執行後會知道結果，如果有得到結果，就印出來。

```
if (result && user) {
   res.write(result);
}
```

目前 index.js:

```
var http = require("http");
const url = require('url'); // 引用內建模組 url
const querystring = require('querystring'); // 引用內建模組 querystring
const GAME = require('./game');
//create a server object:
http
 .createServer(function (req, res) {
   let { query } = url.parse(req.url);
   let { user } = querystring.parse(query);
   let result = GAME.play(user); // 呼叫 GAME.play
   res.writeHead(200, { "Content-Type": "text/html; charset=utf-8" });
   if (result && user) { // 如果 result 跟 user 都有值的話，就印出來。
     res.write(result); //write a response to the client
   } else {
     res.write(' 無資料 ');
   }
   res.end(); //end the response
 })
 .listen(8080); //the server object listens on port 8080
```

6. 試著在網址列後輸入 /?user= 石頭

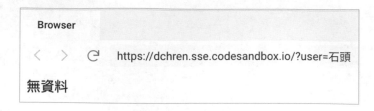

因為 result 沒有得到回傳值，所以是無資料

7. 但你可以試著加

```
console.log(result);
```

印出來值是 undefined

## 6-4 撰寫 GAME 物件

### 6-4-1 宣告常數

1. 先宣告常數也就是剪刀石頭布

```
const ROCK = ' 石頭 ';
const SCISSOR = ' 剪刀 ';
const PAPER = ' 布 ';
```

2. 為了完善一點，將這些常數再開一個 const.js 的檔案放在裡面。

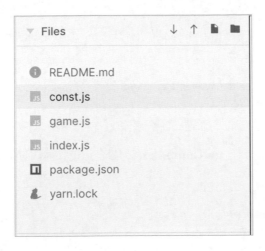

3. 裡面宣告為

const.js:

```
const ROCK = '石頭';
const SCISSOR = '剪刀';
const PAPER = '布';

exports.ROCK = ROCK;
exports.SCISSOR = SCISSOR;
exports.PAPER = PAPER;
```

4. 回到 game.js 將其引用進來

```
const { ROCK, SCISSOR, PAPER } = require('./const');
```

這麼做的好處是當日後有常數需要做更改時，只要到 const.js 去尋找就可以了。

## 6-4-2 撰寫遊戲邏輯

剪刀石頭布的遊戲邏輯就是，將 user 所輸入的值跟電腦輸出的值，兩相比較，看誰勝誰負。

1. 因此在 play 函式裡的第一步先拿電腦出的拳，所以要宣告 getComputer() 函式，可以放在 GAME 物件裡，也可以放外面，但是建議原則上，不會被外部呼叫到的函式，都不要放在物件裡。

```
const getComputer = () => {
 return [ROCK, SCISSOR, PAPER][Math.floor(Math.random() * 3)]; // 取隨機數
乘以 3，再去小數點後，得到 0~2 ，直接從陣列取值回傳
};
```

2. 在 play 函式裡加上 getComputer() 函式，並預期可以獲得回傳電腦出的拳。

```
let com = getComputer();
```

3. 再將 user 跟 com 去做比對，就能獲得猜拳的贏家

```
let winner = getWinner(user,com);
```

4. getWinner() 函式就要寫一下條件判斷式去做比較了

```
const getWinner = (user, com) => {
 if (user === com) {
   // 如果兩者相同就是平手
   return '平手';
 } else if (
   // 玩家贏的情況
   (user === ROCK && com === SCISSOR) ||
   (user === SCISSOR && com === PAPER) ||
   (user === PAPER && com === ROCK)
 ) {
   return '玩家';
 } else {
   // 非上述都是電腦贏
   return '電腦';
 }
};
```

5. play() 函式將結果以物件方式 return 回去

```
return {
    com,
    user,
    winner
};
```

目前 game.js

```
const { ROCK, SCISSOR, PAPER } = require('./const');
const getComputer = () => {
 return [ROCK, SCISSOR, PAPER][Math.floor(Math.random() * 3)];
};

const getWinner = (user, com) => {
 if (user === com) {
    // 如果兩者相同就是平手
    return ' 平手 ';
 } else if (
    // 玩家贏的情況
    (user === ROCK && com === SCISSOR) ||
    (user === SCISSOR && com === PAPER) ||
    (user === PAPER && com === ROCK)
 ) {
    return ' 玩家 ';
 } else {
    // 非上述都是電腦贏
    return ' 電腦 ';
 }
};

const GAME = {
 play: (user) => {
   if (!user) {
      return;
   }

   let com = getComputer();
   let winner = getWinner(user, com);
   return {
```

```
      com,
      user,
      winner
    };
  }
};

module.exports = GAME;
```

6. 因為我們已將回傳值改寫為物件了，所以 index.js 的 res.write(result) 就必須改寫成

```
res.write(
      ' 玩家出 ${result.user}, 電腦出 ${result.com}, 贏家是 ${result.
winner}'
      );
```

目前的 index.js：

```
var http = require("http");
const url = require('url'); // 引用內建模組 url
const querystring = require('querystring'); // 引用內建模組 querystring
const GAME = require('./game');
//create a server object:
http
 .createServer(function (req, res) {
   let { query } = url.parse(req.url);
   let { user } = querystring.parse(query);
   let result = GAME.play(user); // 呼叫 GAME.play
   console.log(result);
   res.writeHead(200, { "Content-Type": "text/html; charset=utf-8" });
   if (result && user) {
     // 如果 result 跟 user 都有值的話，就印出來。
     res.write(
       ' 玩家出 ${result.user}, 電腦出 ${result.com}, 贏家是 ${result.winner}'
     );
   } else {
     res.write(' 無資料 ');
   }
   res.end(); //end the response
 })
 .listen(8080); //the server object listens on port 8080
```

7. 在網址列後輸入 /?user= 布 就會印出

但用手打的方式很不直覺,所以我們在這裡寫一點前端 ( 網頁 ),讓剪刀石頭布,可以直接在網頁上點選。

8. 首先將剪刀石頭布引用進 index.js

```
const { ROCK, SCISSOR, PAPER } = require('./const');
```

然後將其輸出成按鈕形式

```
[ROCK, SCISSOR, PAPER].map((v) => {
    res.write(
      '<input type="button" value="${v}" onclick="location.href='/ ?
user=${v}'" />'
    ); // 以 html 的 button 形式寫出去
  });
```

9. 一開始會是

10. 點擊後，便會輸出結果

目前 index.js :

```javascript
var http = require("http");
const url = require('url'); // 引用內建模組 url
const querystring = require('querystring'); // 引用內建模組 querystring
const GAME = require('./game');
const { ROCK, SCISSOR, PAPER } = require('./const');

//create a server object:
http
 .createServer(function (req, res) {
  let { query } = url.parse(req.url);
  let { user } = querystring.parse(query);
  let result = GAME.play(user); // 呼叫 GAME.play
  console.log(result);
  res.writeHead(200, { "Content-Type": "text/html; charset=utf-8" });
  [ROCK, SCISSOR, PAPER].map((v) => {
    res.write(
      '<input type="button" value="${v}" onclick="location.href='/ ?
user=${v}'" />'
    ); // 以 html 的 button 形式寫出去
  });

  if (result && user) {
    // 如果 result 跟 user 都有值的話，就印出來。
    res.write(
      ' 玩家出 ${result.user}, 電腦出 ${result.com}, 贏家是 ${result.winner}'
    );
  } else {
    res.write(' 無資料 ');
  }
  res.end(); //end the response
})
 .listen(8080); //the server object listens on port 8080
```

### 6-4-3 紀錄遊戲的輸贏：探討 閉包 Clourse

1. 第一個想法就是在物件上宣告一個 history 的空陣列

```
history: []
```

2. 將每一次的結果 push 進去

```
GAME.history.push(result);
```

3. 宣告 getHistory() 來取用歷史資料。

```
getHistory: () => {
   return GAME.history;
}
```

目前 game.js 裡的 GAME 物件為：

```
const GAME = {
 history: [],
 play: (user) => {
   if (!user) {
     return;
   }

   let com = getComputer();
   let winner = getWinner(user, com);
   let result = {
     com,
     user,
     winner
   };
   GAME.history.push(result); // 將結果 push 進 GAME.history
   return result;
 },
 getHistory: () => {
   return GAME.history;
 }
};
```

4. 在 index.js 裡將記錄印出來

```
GAME.getHistory().map((v, i) => {
    res.write(
      '<div> 第 ${i + 1} 把：玩家出 ${v.user}, 電腦出 ${v.com}, 贏家是 ${
        v.winner
      }</div>'
    );
});
res.end(); //end the response
```

目前的 index.js：

```
var http = require("http");
const url = require('url'); // 引用內建模組 url
const querystring = require('querystring'); // 引用內建模組 querystring
const GAME = require('./game');

const { ROCK, SCISSOR, PAPER } = require('./const');


//create a server object:
http
 .createServer(function (req, res) {
   let { query } = url.parse(req.url);
   let { user } = querystring.parse(query);
   let result = GAME.play(user); // 呼叫 GAME.play
   console.log(GAME.getHistory());
   res.writeHead(200, { "Content-Type": "text/html; charset=utf-8" });
   [ROCK, SCISSOR, PAPER].map((v) => {
     res.write(
       '<input type="button" value="${v}" onclick="location.href='/ ?
user=${v}'" />'
     ); // 以 html 的 button 形式寫出去
   });


   if (result && user) {
     // 如果 result 跟 user 都有值的話，就印出來。
     res.write(
       ' 玩家出 ${result.user}, 電腦出 ${result.com}, 贏家是 ${result.winner}'
```

```
    );
  } else {
    res.write(' 無資料 ');
  }

  GAME.getHistory().map((v, i) => {
    res.write(
      '<div> 第 ${i + 1} 把：玩家出 ${v.user}, 電腦出 ${v.com}, 贏家是 ${
        v.winner
      }</div>'
    );
  });
  res.end(); //end the response
})
.listen(8080); //the server object listens on port 8080
```

5. 這雖然可以運作

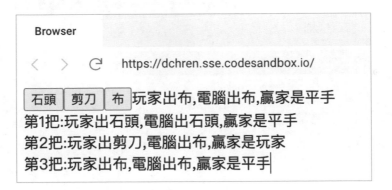

6. 但是在 index.js 卻可以叫出 GAME.history ；

我們只希望 index.js 知道 GAME.getHistory 函式就好，不要能呼叫使用
GAME.history 這個變數，避免變數遭到外部檔案的修改。

有兩個作法：

## \* 方法一

直接將 history 宣告在 GAME 物件外

```
const history = [];
```

目前 game.js 裡的 GAME 物件為：

```
const history = []; // 宣告在外面
const GAME = {
 //history: [], 註解掉
 play: (user) => {
   if (!user) {
     return;
   }

   let com = getComputer();
   let winner = getWinner(user, com);
   let result = {
     com,
     user,
     winner
   };
   history.push(result); // 改成將結果 push 進 history
   return result;
 },
 getHistory: () => {
   return history; // 改成 history
 }
};

module.exports = GAME;
```

這雖然可以避免外部檔案不會去引用到內部參數，但難保日後撰寫 game.
js 的新功能時，不小心使用到 history 參數，造成不必要的錯誤，進而產生
bug。

## * 方法二：閉包 clourse

1. 將 GAME 從物件改成箭頭函式 getGameObject() 函式，最後 return 我們原先的 GAME 資料結構出去

```
const getGameObject = ()=>{
  return {
    play: ...
  }
}
module.exports = getGameObject();
```

2. 將 history 放在函式裡面

```
const getGameObject = () => {
 let history = [];
…
```

目前的 getGameObject：

```
const getGameObject = () => {
 let history = [];
 return {
   play: (user) => {
     if (!user) {
       return;
     }

     let com = getComputer();
     let winner = getWinner(user, com);
     let result = {
       com,
       user,
       winner
     };
     history.push(result); // 將結果 push 進 history
     return result;
   },
   getHistory: () => {
     return history;
   }
 };
};
```

index.js 就無法叫用 GAME.history 了。

game.js 也無法去呼叫到 GAME.history，因為 history 已經被包在 getGameObject() 函式裡了。

這樣做的好處，就可以避免在 game.js 裡撰寫其他程式時，誤去修改到 history 的值，其他如果有類似需求的變數或函式也可以如法泡製，例如 getComputer() 跟 getWinner()。

目前的 game.js:

```js
const { ROCK, SCISSOR, PAPER } = require('./const');

const getGameObject = () => {
 let history = [];
 let getComputer = () => {
   return [ROCK, SCISSOR, PAPER][Math.floor(Math.random() * 3)];
 };
 let getWinner = (user, com) => {
   if (user === com) {
     // 如果兩者相同就是平手
     return '平手';
   } else if (
     // 玩家贏的情況
     (user === ROCK && com === SCISSOR) ||
     (user === SCISSOR && com === PAPER) ||
     (user === PAPER && com === ROCK)
   ) {
     return '玩家';
   } else {
     // 非上述都是電腦贏
     return '電腦';
   }
 };
};
```

```
  return {
    play: (user) => {
      if (!user) {
        return;
      }
      let com = getComputer();
      let winner = getWinner(user, com);
      let result = {
        com,
        user,
        winner
      };
      history.push(result);
      return result;
    },
    getHistory: () => {
      return history;
    }
  };
};

module.exports = getGameObject();
```

目前的 index.js:

```javascript
var http = require("http");
const url = require('url'); // 引用內建模組 url
const querystring = require('querystring'); // 引用內建模組 querystring
const GAME = require('./game');
const { ROCK, SCISSOR, PAPER } = require('./const');

//create a server object:

http
  .createServer(function (req, res) {
    let { query } = url.parse(req.url);
    let { user } = querystring.parse(query);

    let result = GAME.play(user);
    res.writeHead(200, { "Content-Type": "text/html; charset=utf-8" });
    [ROCK, SCISSOR, PAPER].map((v) => {
      res.write(

        '<input type="button" value="${v}" onclick="location.
href='/ ? user=${v}'" />'

      );
    });

    if (result && user) {
      res.write(
        ' 玩家出 ${result.user}, 電腦出 ${result.com}, 贏家是 ${result.
winner}'
      );
      //write a response to the client
    } else {
      res.write(' 無資料 ');
    }
    GAME.getHistory().map((v, i) => {
      res.write(
        '<div> 第 ${i + 1} 把 : 玩家出 ${v.user}, 電腦出 ${v.com}, 贏家是 ${
          v.winner
        }</div>'
      );
    });
    res.end(); //end the response
  })
  .listen(8080); //the server object listens on port 8080
```

```
Browser
< > C  https://dchren.sse.codesandbox.io/

石頭  剪刀  布 玩家出布,電腦出布,贏家是平手
第1把:玩家出石頭,電腦出石頭,贏家是平手
第2把:玩家出剪刀,電腦出布,贏家是玩家
第3把:玩家出布,電腦出布,贏家是平手
```

## 6-4-4　class vs clourse 想一想

實現一個 Cat 物件可以用 class 和 clourse

class:

```
class Cat{
  #name; // 宣告為內部屬性，外部無法直接呼叫
  constructor(n){
    this.#name = n;
  }
  run(){ console.log('${this.#name} 跑步中 ')}
}
let cat1 = new Cat('cathy');
cat1.run(); //cathy 跑步中
```

clourse:

```
const getCat = (n) => {
  let name = n;
  return {
    run() {
      console.log('${name} 跑步中 ');
    }
  };
};
let cat2 = getCat('cathy');
cat2.run(); //cathy 跑步中
```

class 和 clourse 都可以實現出一樣的特性，想一想你比較喜歡那一個呢？

## ▌ 6-5 遊戲邏輯程式碼放在前端

JS 原本就是前端的語法,那我們能不能直接將 game.js 放在前端呢?

當然可以囉,我們試著將程式搬家到前端。

1. 首先,我們不修改原本的進入點 index.js 檔案

2. 另外新增一個檔案 main.js 來做為進入點

main.js:

```javascript
var http = require("http"),
 url = require("url"),
 path = require("path"),
 fs = require("fs");

console.log('into main.js');

http
 .createServer(async function (req, res) {
   var uri = url.parse(req.url).pathname;
   res.writeHead(200, {
     "Content-Type": 'text/${
       uri.search('.js') > -1 ? 'javascript' : 'html'
     }; charset=utf-8'
   });
   let file = path.join(__dirname, uri);
   let hasFile = await fs.existsSync(file);
   if (hasFile) {
     let fileStream = fs.createReadStream(file);
     fileStream.pipe(res);
   } else {
     res.writeHead(200, { "Content-Type": "text/plain" });
     res.write("404 Not Found\n");
     res.end();
   }
 })
 .listen(8080); //the server object listens on port 8080
```

3. 修改 package.json:

```
"main": "main.js",
"scripts": {
  "start": "nodemon main.js"
},
```

4. 新增 index.html

```html
<html>
 <head>
   <title> 剪刀石頭布 </title>
   <script src="/const.js"></script>
   <!-- 將檔案 const.js 引用進來 -->
   <script src="/game.js"></script>
   <!-- 將檔案 game.js 引用進來 -->

   <script>
     let GAME = getGameObject(); // 呼叫 getGameObject 函式，產生 GAME 物件
     let play = (v) => {
       // 定義 play 函式
       let result = GAME.play(v); // 呼叫 GAME.play 函式
       console.log(result);
       let result_div = document.getElementById('result'); // 操作 DOM
       result_div.innerText = ' 玩家出 ${result.user}, 電腦出 ${result.com}, 贏家
是 ${result.winner}'; // / 印出結果
       let divs = GAME.getHistory().map((v, i) => {
         // 獲得歷史紀錄
         return '<div> 第 ${i + 1} 把 : 玩家出 ${v.user}, 電腦出 ${v.com}, 贏家是 ${
           v.winner
         }</div>';
       });
       let history_div = document.getElementById('history'); // 操作 DOM
       history_div.innerHTML = divs.join(''); // 印出歷史紀錄
     };
   </script>
 </head>
 <body>
   <h1> 剪刀石頭布 </h1>
   <input type="button" value=" 剪刀 " onclick="play(' 剪刀 ')" />
   <!-- 呼叫 play 函式，參數為 剪刀 -->
   <input type="button" value=" 石頭 " onclick="play(' 石頭 ')" />
   <!-- 呼叫 play 函式，參數為 石頭 -->
```

```
<input type="button" value="布" onclick="play('布')" />
<!-- 呼叫 play 函式，參數為 布 -->
<div id="result"></div>
<div id="history"></div>
</body>
</html>
```

5. 然後 restart server

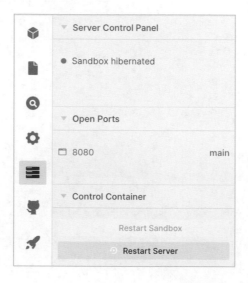

6. Browser 加網址 /index.html 開啟 ( 沒加的話，會因為找不到檔案而產生錯誤。)

7. 會看到這個畫面，但按了不會有任何動作

8. 先按最右邊的 <span></span>，直接開啟一個分頁

9. 將 console 打開，會看到報錯 Uncaught ReferenceError: exports is not defined 等

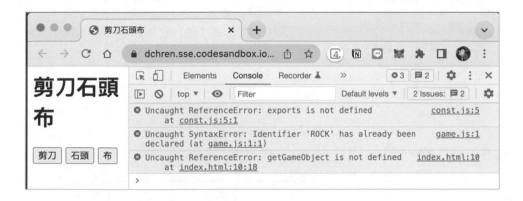

這是因為 exports , require 的語法在前端 JS ，也就是 Browser 不支援。

10. 到 codesandbox 裡，將有用到 exports 的部份，先註解掉並存檔

const.js:

```
const ROCK = '石頭';
const SCISSOR = '剪刀';
const PAPER = '布';

// exports.ROCK = ROCK;
// exports.SCISSOR = SCISSOR;
// exports.PAPER = PAPER;
```

game.js:

```
//const { ROCK, SCISSOR, PAPER } = require('./const');
```

```
//module.exports = getGameObject();
```

11. 重新整理 browser ，就可以看到結果，跟 console.log 印出的內容

## 6-5-1 觀察一下，程式放在前後端的差別

| | 前端 | 後端 |
|---|---|---|
| 優點 | 降低後端運算需求 | 減少前端不確定性<br>隱藏程式碼 |
| 缺點 | 明碼 | 增加後端運算需求 |

至於好壞就是自己撰寫程式取捨了。

另外關於前端 JS 也就是 Browser 還不支援的語法，其實是可以用 webpack 這類的打包模組去做 JS 程式轉譯的動作，將不支援的語法給轉譯成新的檔案，就不需要手動註解了。

參考：https://webpack.docschina.org/guides/getting-started/

## 6-5-2 避免巢狀結構

通常遊玩頁，跟歷史紀錄應該是要分開不同的網址去呼叫程式，但在原本 http 的模組裡，得藉由 url.parse(req.url); 去處理 path 的部份，程式就會長成：

```
if (path === '/') {
  if (result && user) {
    ...
  }
}else if (path === '/history') {
  ...
}
```

if 裡面還有 if，也就是條件式裡面還有條件式，這就是巢狀結構。

巢狀結構很容易產生錯誤，應該要避免掉，但原始的 http 模組很難避免掉寫出巢狀結構。

那之於上述的理由，以及其他的問題，就有開發者為了改進這些問題，開發出了網站框架，只要依循著框架的開發邏輯，就可以快速開發出所需要的應用。

## 6-6 使用 framework 框架 express

首先安裝 express，不同於之前所以安裝的 nodemon 是 tool 工具， url, querystring 是 module 模組，express 是 framework 框架。

框架的意思是，我們只要依照著其結構及開發邏輯，就可以快速的開發出我們想要的應用。

| 名稱 | 說明 | js 知名套件 |
|---|---|---|
| tool 工具 | 特定功能。 | nodemon, webpack |
| module 模組 | 某特定領域的函式集成，可以供呼叫引用。 | http , url , querystring , jquery |
| framework 框架 | 依照其指引開發某特定領域的應用，例如：開發前端網頁。<br>通常範例 code 會有 hello world 跟 todo list | express , react , vue , angular , botframework |

例如 node.js 開發網站要用到 http，url，querystring 等等的 module，但是藉由 express 來開發，就不需要去一個一個引用到上述的 module，開發網站上省時省力很多，但是 http 等等的 module 不是不見了，而是被寫在express 裡面了。

例如：https://github.com/expressjs/express/blob/158a17031a2668269aedb31ea07b58d6b700272b/lib/application.js 的就有引用到 http 等。

```
/*!
 * express
 * Copyright(c) 2009-2013 TJ Holowaychuk
 * Copyright(c) 2013 Roman Shtylman
 * Copyright(c) 2014-2015 Douglas Christopher Wilson
 * MIT Licensed
 */

'use strict';

/**
 * Module dependencies.
 * @private
 */

var finalhandler = require('finalhandler');
var Router = require('./router');
var methods = require('methods');
var middleware = require('./middleware/init');
var query = require('./middleware/query');
var debug = require('debug')('express:application');
var View = require('./view');
var http = require('http'); // 這裡引用了 http
var compileETag = require('./utils').compileETag;
var compileQueryParser = require('./utils').compileQueryParser;
var compileTrust = require('./utils').compileTrust;
var deprecate = require('depd')('express');
var flatten = require('array-flatten');
var merge = require('utils-merge');
var resolve = require('path').resolve;
var setPrototypeOf = require('setprototypeof')
```

## 6-6-1 Hello World

我們延續原來的 codesandbox

1. 先至 Dependencies 下的 Add dependency 安裝 express

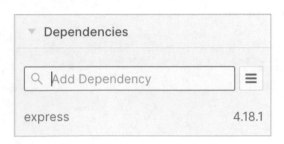

2. 我們另外開一個新檔案 app.js 來寫 express 的 code

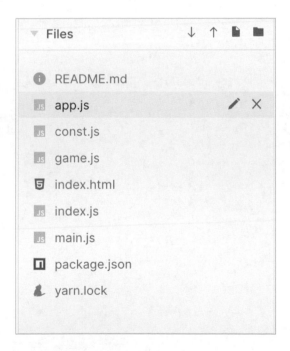

3. 編輯 app.js，先宣告引用 express

```
const express = require('express');
```

4. 藉由執行 express() 產生 app 物件

```
const app = express();
```

5. 寫一個最基本的路由，當 browser 訪問 根目錄 '/' 或 路由 route '/' 時，
印出 'hello world'

```
app.get('/', (req, rep) => { // browser 訪問 根路由時
 rep.end('hello world'); // 回應 'hello world'
});
```

6. 監聽 8080 port

```
app.listen(8080, () => {
 console.log('listening');
});
```

目前 app.js:

```
const express = require('express');
const app = express();

app.get('/', (req, rep) => {
 rep.end('hello world');
});

app.listen(8080, () => {
 console.log('listening');
});
```

7. 修改 package.json

```
"main": "app.js",
"scripts": {
  "start": "nodemon app.js"
},
```

8.　重啟 Restart Server

9.　重新整理 Browser 就會印出 hello world 了

10.　試著讓 browser 訪問其他網址

會回應找不到 cannot GET/a，這裡和 http 的 createServer 函式的回應，明顯是不同的。

而這個 /xxx/xxx/ ⋯ 等的網址，我們稱做為 路由 routing

## 6-6-2 路由 route

在後端應用網站等開發上,路由是一個相當重要的議題,路由可以有很多作法,不管是前端路由管理,或是後端路由作分散式架構等,背後牽扯到的設計也非常廣泛,不管是從前端來看,還是後端來看,都是相當精采。

在 express 的 route ,就是將 http 裡,使用者訪問時 ,http 協議所使用的定義的 get post delete put 等方法,來做為路由的定義。

app.METHOD(PATH, HANDLER)

- METHOD 可以替換成 get post delete put 等
- PATH 就是路徑,第一個通常是根目錄 '/'
- HANDLER 就是對應的函式,例如

```
app.get('/', (req, rep) => {
 rep.end('hello world');
});
```

其中的

```
(req, rep) => {
 rep.end('hello world');
})
```

就是當使用者訪問時,會由這個函式,來做處理。

## 6-6-3 使用 express 改寫 剪刀 石頭 布

為了了解使用框架開發的好處,我們將前面章節所示範的前後端部份改寫。

### * 使用 express 作前端的撰寫

1. 在 app.js 加上

```
app.use(express.static(path.join(__dirname)));
```

直接去讀取 / 根目錄下的檔案

目前 app.js：

```
const express = require('express');
const path = require("path"); // 引用 path
const app = express();
app.use(express.static(path.join(__dirname))); // 直接開放讀取 / 目錄下的
檔案
app.listen(8080, () => {
 console.log('listening');
});
```

2.　重新整理 Browser

3.　可以玩了

對比一下前面章節，為了要讓 main.js 順利讀取檔案並印出內容，就寫了 20
行，而 express 一行搞定 。

＊　使用 **express** 作後端的撰寫

1.　先將先前註解掉的語法加回來

const.js:

```
const ROCK = '石頭';
const SCISSOR = '剪刀';
const PAPER = '布';

exports.ROCK = ROCK;
exports.SCISSOR = SCISSOR;
exports.PAPER = PAPER;
```

game.js:

```
const { ROCK, SCISSOR, PAPER } = require('./const');

module.exports = getGameObject();
```

2.  註解掉 `// app.use(express.static(path.join(__dirname)));`
    避免衝突

3.  拿 query 值 加上

```
let { user } = req.query;
```

4.  跟 http 模組一樣需要寫入檔頭 head 告訴來訪問的瀏覽器，這個伺服器
    將用 utf-8 輸出

```
res.writeHead(200, { "Content-Type": "text/html; charset=utf-8" });
```

5.  其餘部份則一樣。

目前 app.js

```javascript
const express = require('express');
const path = require("path"); // 引用 path
const app = express();
const GAME = require('./game');
const { ROCK, SCISSOR, PAPER } = require('./const');
// app.use(express.static(path.join(__dirname))); // 先註解掉，避免衝突

app.get('/', (req, res) => {
 let { user } = req.query; // 拿 query 值，只要一行
 let result = GAME.play(user);
 res.writeHead(200, { "Content-Type": "text/html; charset=utf-8" });
 [ROCK, SCISSOR, PAPER].map((v) => {
   res.write(
     '<input type="button" value="${v}" onclick="location.href='/ ?
user=${v}'" />'
   );
 });
 if (result && user) {
   res.write(
     ' 玩家出 ${result.user}, 電腦出 ${result.com}, 贏家是 ${result.winner}'
   );
 } else {
   res.write(' 無資料 ');
 }
 GAME.getHistory().map((v, i) => {
   res.write(
     '<div> 第 ${i + 1} 把 : 玩家出 ${v.user}, 電腦出 ${v.com}, 贏家是 ${v.
winner}</div>'
   );
 });
 res.end();
});

app.listen(8080, () => {
 console.log('listening');
});
```

6. 重新整理 Browser

7. 可以玩了

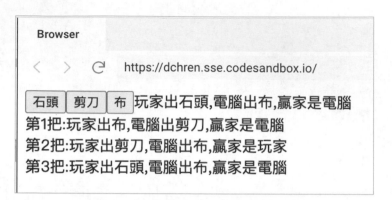

8. 如果要寫一個新的路由，只需要往下加，就可以了

例如：我要新增一個 API 用來獲得 遊戲的歷史資料

```
app.get('/history', (req, res) => {
 res.json(GAME.getHistory());
});
```

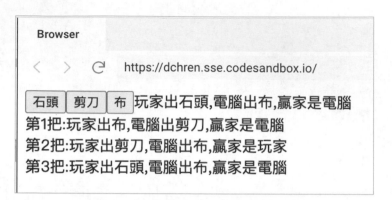

## 6-6-4 練習

電腦的運算判斷速度，通常 boolean 布林值最快，數字次之，字串最慢，所以前面的宣告

```
const ROCK = '石頭';
const SCISSOR = '剪刀';
const PAPER = '布';
```

可以改成

```
const ROCK = 1;
const SCISSOR = 2;
const PAPER = 3;
```

而 game.js 的 getWinner 的判斷，就可以改成數字運算

例如 user - com === 0 就是平手

試著將本章用數字運算的方式，重頭改寫練習一次

## ▍附帶一提：使用 webpack 打包

前面章節將 const.js 及 game.js 置於前端時，必須將前端瀏覽器尚未支援的 export 等關鍵字註解掉，但其實有更好的作法，就是利用 webpack 來做程式碼的打包及轉譯。

1. 安裝 webpack , webpack-cli 並在 package..json 移到 devDependencies

```
"devDependencies": {
    "@types/node": "^17.0.21",
    "webpack": "5.74.0",
    "webpack-cli": "4.10.0"
 }
```

2. 改寫目前的 app.js

```
const express = require('express');
const path = require("path"); // 引用 path
const app = express();
app.use(express.static(path.join(__dirname)));

app.listen(8080, () => {
 console.log('listening');
});
```

3. game.js 的 getGameObject 將其新增為 window 全域物件下的內建函式

```
window.getGameObject = getGameObject;
```

4. 新增 webpack.config.js

```
module.exports = {
 entry: ["./const.js", "./game.js"],
 output: {
   filename: "bundle.js"
 }
};
```

5. package.json 的 scripts 新增 build

```
"scripts": {
  "start": "node app.js",
  "build": "webpack"
},
```

6. 下指令 打包 const.js game.js

```
npm run build
```

7. 好了之後，會多一個 bundle.js 的檔案

8. 註解掉之前的引用，將 bundle.js 引用進 index.html

```
<!-- <script src="/const.js"></script> -->
<!-- <script src="/game.js"></script> -->
<script src="/dist/bundle.js"></script>
```

9. 改寫 index.html 的 script 的部份，改成從 window 全域物件下呼叫 getGameObject

```
let GAME = window.getGameObject();
```

10. 重新整理 browser 試試

參考：https://webpack.docschina.org/guides/getting-started/

## 說明

這個簡單示範 webpack 打包成 bundle.js 後，將 exports 的語法轉譯為 browser 可以接受的語法。

關於 webpack 的設定，其實還有很多，如對 webpack 有興趣，諸如原理、設定等，請參考 https://webpack.docschina.org/guides/getting-started/

# 7

# 開源篇
# npm & github

## ▋ 7-1　學習目標／演練成果

藉由實作 npm 套件，使了解開放性社群如何能促進 IT 產業的進步。

## ▋ 7-2　實做開源套件：天氣小幫手

中央氣象局的 open data 相當準確，但是回傳回來的資料欄位相當複雜，我們以此為例，將其修改為只回傳當下的天氣狀況，並且發佈為新的套件天氣小幫手，供開發者社群使用。

### 7-2-1　申請中央氣象局資料開放平台

1.　先 google 中央氣象局資料開放平台

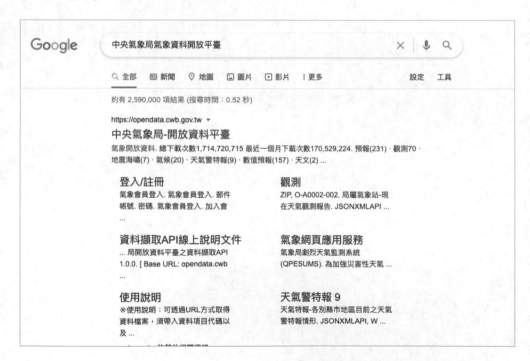

2. 進來之後，先點 登入 / 註冊

   https://opendata.cwb.gov.tw/devManual/insrtuction

3. 註冊一個新帳號

4. 點 API 授權碼
5. 取得授權碼

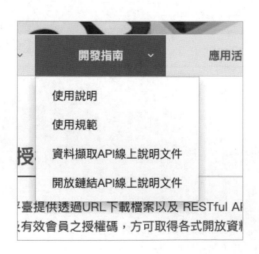

## API授權碼

本平臺提供透過URL下載檔案以及 RESTfu
以及有效會員之授權碼，方可取得各式開放

**一、取得授權碼**

會員之授權碼可於下方按鈕取得

取得授權碼

**二、更新授權碼**

一旦更新授權碼後，舊的授權碼將永久失效

更新授權碼

6.　獲得 授權碼 後

7.　複製 授權碼 。

8.　點 開發指南

開發指南 ∨　　應用活

使用說明

使用規範

資料擷取API線上說明文件

開放鏈結API線上說明文件

平臺提供透過URL下載檔案以及 RESTful AI
有效會員之授權碼，方可取得各式開放資

9. 點 資料擷取 API 線上說明文件

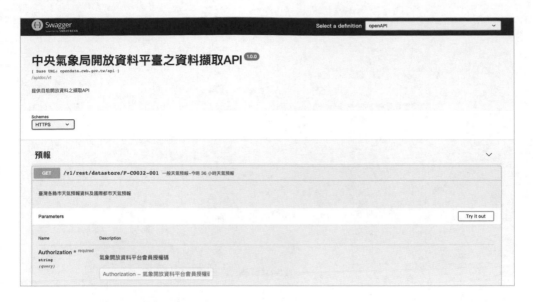

10. 另開新頁後

11. 打開 第一個 一般天氣預報 - 今明 36 小時天氣預報

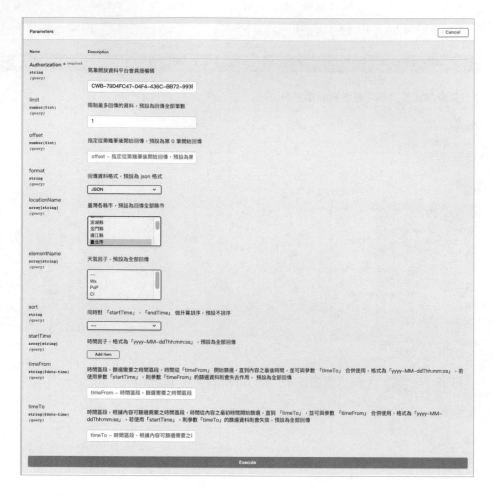

12. 點 Try it out

13. 貼上 授權碼 在 Authorization 欄位

14. 在 locationName 選一個縣市

15. 點 Execute

在最下面 Response 回來的 Data 會看到回傳回來的資料欄位很複雜，但我們只想要當下的天氣狀況而已。

所以我們來改動這個部份，並且發佈新的套件，供社群使用。

## 7-2-2 撰寫

1. 至 codesandbox 開 node http server ，安裝完 nodemon 且改寫 package. json 裡的 start 為 nodemon index.js

2. 新增一個檔案 TaiwanWeather.js 作為我們撰寫邏輯程式的地方

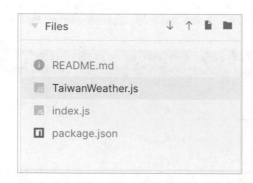

3. 我們需 node-fetch 套件來協助我們取中央氣象局 open data 的 API

版本號，請選 2.6.7 ；新版和 codesandbox 的預設 node.js 環境不相同，會無法運行。

4. 先試一下 API 能否存取

```
const fetch = require('node-fetch'); // 引用 node-fetch
(async () => { // 因為 fetch 是 promise 的函式，所以要用 async
 let url = ''; //url 的部份，將剛剛的 Request URL 貼過來
 let response = await fetch(url); // 因為 fetch 是 promise 的函式，所以要用
await
})();
```

5.　url 的部份，將剛剛的 Request URL 貼過來

目前的 TaiwanWeather.js：

```
const fetch = require('node-fetch');
(async () => {
  let url = 'https://opendata.cwb.gov.tw/api/v1/rest/datastore/F-C0032-
001 ? ...';
  let response = await fetch(url);
  let data = await response.json(); // response 的資料還需轉譯成 json 格式
  console.log(data); // 印出資料
})();
```

6.　開一個新 terminal，點 +

7.　下 node TaiwanWeather.js

可以看到欄位裡還有資料，造成取用資料時的不便

8.　我們先取出當下的天氣狀況的資料

```
let { weatherElement } = data.records.location[0];
let [Wx, PoP, MinT, CI, MaxT] = weatherElement.map((i) => {
  return i.time[0].parameter.parameterName;
});
console.log(Wx, PoP, MinT, CI, MaxT);
```

9.　再下一次 node TaiwanWeather.js　印出來看看

```
sandbox@sse-sandbox-jxkeyu:/sandbox$ node TaiwanWeather.js
晴午後短暫雷陣雨 80 25 舒適至悶熱 34
```

10. Wx, PoP, MinT, CI, MaxT 這些就是我們要的資料

目前的 TaiwanWeather.js：

```javascript
const fetch = require('node-fetch');
(async () => {
 let url = 'https://opendata.cwb.gov.tw/api/v1/rest/datastore/F-C0032-
001 ? Authorization=CWB-7DC207A4-3A99-4610-8E42-A2371E023ADB';
 let response = await fetch(url);
 let data = await response.json(); // response 的資料還需轉譯成 json 格式
 // console.log(data); // 印出資料
 let { weatherElement } = data.records.location[0];
 let [Wx, PoP, MinT, CI, MaxT] = weatherElement.map((i) => {
   return i.time[0].parameter.parameterName;
 });
 console.log(Wx, PoP, MinT, CI, MaxT);
})();
```

# 7-3 改寫成可以發佈的形式

在 clourse 還沒成為主流之前，會是用 class 來寫，比較工整；不過在目前，都會使用 clourse 的方式來寫，程式會比較簡潔。

1. 首先在 TaiwanWeather.js 下面 export 函式出去

```javascript
module.exports = ()=>{
}
```

2. 將前一節的函式引入並改寫

3. 將授權碼改成變數 key 及加入 locationName 的參數，讓使用者可以輸入城市

```javascript
let url = 'https://opendata.cwb.gov.tw/api/v1/rest/datastore/F-C0032-001 ?
Authorization=${key}&locationName=${encodeURI(
  city
)}'; // 用 encodeURI 將中文編碼
```

4. 故輸出函式加入輸入參數 key 及 city 其餘內容一樣

```
module.exports = async (key, city)
```

目前的 TaiwanWeather.js：

```
const fetch = require('node-fetch');
module.exports = async (key, city) => { // key 值就是 授權碼 , city 則是要
查那個城市，只能用中央氣象局的格式，例如：臺北市，詳資料擷取 API 線上說明文件的
locationName
 let url = 'https://opendata.cwb.gov.tw/api/v1/rest/datastore/F-C0032-
001 ? Authorization=${key}&locationName=${encodeURI(
   city
 )}'; // 用 encodeURI 將中文編碼
 let response = await fetch(url);
 let data = await response.json();
 let { weatherElement } = data.records.location[0];
 let [Wx, PoP, MinT, CI, MaxT] = weatherElement.map((i) => {
   return i.time[0].parameter.parameterName;
 });
 return {
   city,
   degree: {
     min: MinT,
     max: MaxT
   },
   weather: Wx,
   feeling: CI,
   chance: PoP
 };
};
```

授權碼改成用輸入的，不再置入自己的。

這是因為一旦發佈出去之後，有用這個套件的開發者，都用你的授權碼去
要氣象局的資料，而你不會希望全部的人都可以用的。

5.　測試一下，到 index.js 引用進來，並且呼叫之

```
const TaiwanWeather = require('./TaiwanWeather.js');
(async () => {
  let data = await TaiwanWeather(process.env.KEY, '新北市');
  console.log(data);
})();
```

6.　在 Secret Keys 輸入氣象局的授權碼

7.　重啟程式，看有無印出資料

好了之後，我們就要將其發佈上 NPM

目前的 index.js：

```
var http = require("http");

//create a server object:
http
 .createServer(function (req, res) {
   res.write("Hello World!"); //write a response to the client
   res.end(); //end the response
 })
 .listen(8080); //the server object listens on port 8080
const TaiwanWeather = require('./TaiwanWeather.js');
(async () => {
 let data = await TaiwanWeather(process.env.KEY, '新北市');
 console.log(data);
})();
```

## 7-3-1 發佈 NPM 套件

1. 將 index.js 內容全部槓掉

   改成

```
module.exports = require("./TaiwanWeather.js");
```

2. 改寫 package.json

```
{
 "name": "taiwan-weather-xxx", // 套件名稱，可以改寫成你自己喜歡的名稱，但不
能和 NPM 上面的套件重覆
 "version": "1.0.0", // 版本
 "main": "TaiwanWeather.js",// 程式進入點
 "license": "MIT",
 "dependencies": {
   "node-fetch": "2.6.7"
 },
 "devDependencies": {
   "@types/node": "^17.0.21"
 }
}
```

3. 改寫 README.md

當套件發佈上線於 npm 或是 github 之後，README.md 這個檔案會被預設
為首頁，所以文件的內容就會是以介紹此套件為主。

md 檔也就是 markdown 檔，有自己的撰寫格式

關於 markdown 的撰寫，可參考 https://markdown.tw/

README.md:

```
# taiwan-weather-xxx //:  口口口口
## Installation //: 口口口口

'''bash
npm install --save taiwan-weather-xxx
'''
## Requirements
首先申請中央氣象局 open data 的帳號並得授權碼

## Sample Code
'''js
const TaiwanWeather = require('xxx');
(async () => {
let data = await TaiwanWeather( 授權碼 , ' 新北市 '); //' 新北市 ' 可以換成其他
中央氣象局授受的城市名稱

console.log(data);
})();
'''

# License
The MIT license
```

4. 註冊 npm 帳號：在 terminal 下

```
npm adduser
```

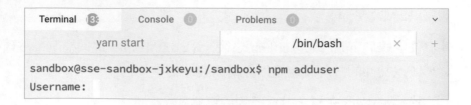

依照著提示，完成註冊。

5. 註冊完成後，就可以發佈了：在 terminal 下

```
npm publish
```

6. 到 npm 網站 www.npmjs.com 點進去

7. 登入前述下 npm add user 時，所註冊的帳號

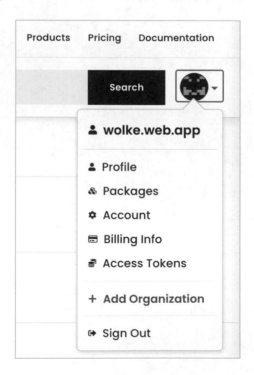

8.　點 packages

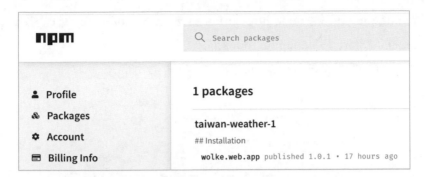

9.　就可以看到剛剛發佈的 package 了

10.　點進去 package

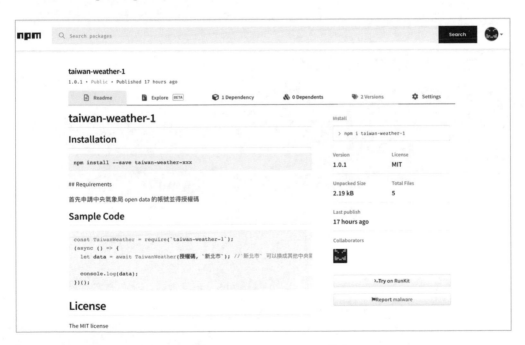

11.　這就是全部的人會看到的樣子

## 7-3-2 測試剛剛發佈的套件

1. 至 codesandbox 開 node http server ，安裝完 nodemon 且改寫 package. json 裡的 start 為 nodemon index.js

2. 因為套件剛發佈 去 Dependencies 那裡搜尋可能會找不到

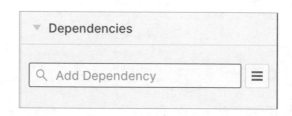

3. 可以直接在 terminal 下指令

```
npm install --save taiwan-weather-xxx
```

如果無法還是無法安裝的話，稍等一下

4. 先新增 Secret Keys ，將授權碼設成 KEY

5. 在 indcx.js 引用進來測試

```
var http = require("http");
const TaiwanWeather = require('taiwan-weather-xxx'); // 引用套件
(async () => {
 let data = await TaiwanWeather(process.env.KEY, '臺南市 '); // 引用 KEY，
並指定查詢城市
 console.log(data);
})();

//create a server object:
http
 .createServer(function (req, res) {
   res.write("Hello World!"); //write a response to the client
   res.end(); //end the response
 })
 .listen(8080); //the server object listens on port 8080
```

6. 重啟 restart sandbox

7. 印出資料，完成

## 7-4 GitHub 發佈

目前我們將套件發佈上 NPM 了，但是程式碼只在我們這裡，如果希望各路大神來幫忙看看 debug ，勢必要將 source code 放在一個開放而且可以協同工作的地方，而這個網站的首選就是 github。

不過目前 codesandbox 對於 github 的支援相當陽春，但以教學來說，還算夠用。

### 7-4-1 註冊 github

1. https://github.com/ 點進去

2. 點 Sign up

3. 依序輸入 email 等資料，驗證等完成註冊

## 7-4-2 上傳 github

1. 點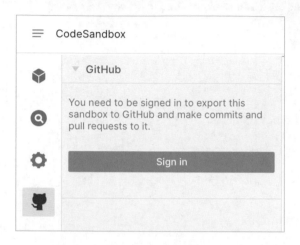

2. 如果是用 google 登入，尚未用 github 帳號登入，會要求你登入 github

3. sign in 你的 github 帳號

4. repository name 輸入 repository 名稱 通常會是套件名稱 taiwan-weather-xxx

5. 點 Create new repository on GitHub

這裡如果出現這個錯誤，先不用擔心，這是 codesandbox 的 bug

開發環境 Environment 被誤認為是開發前端 create-react-app，可以先按上一頁，就會回到正常狀態了。

6. 到 github 選 your repositories

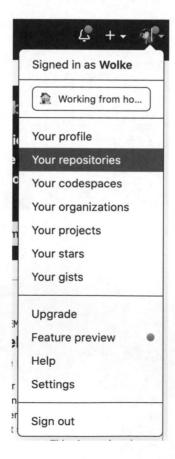

7.　就會看到剛剛上傳的 repo 了

8.　點進去，就是大家會看到的樣子

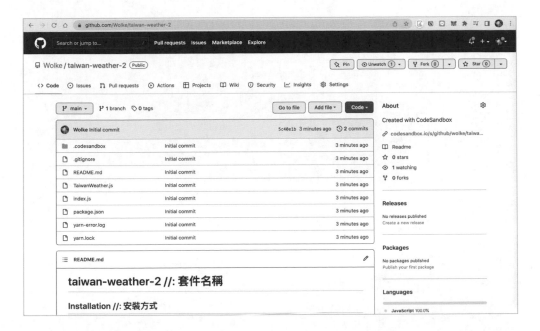

## 7-5 加一個 nodeman 避免 codesandbox 錯認 為 前端開發環境

如出現前述的錯誤，主因是 codesandbox 將後端開發環境跟前端開發環境的開發介面是不同的，所以 codesandbox 誤認該 package 是開發前端環境。

Environment 寫的是 create-react-app

為了解決這個問題，只要裝好 nodemon 這個套件，就可以讓 codesandbox 抓到關鍵套件，正確進入 codesandbox 的後端開發環境。

1.　先安裝好 nodemon

2.　編輯 package.json，將 nodemon 從 dependencies 移至 devDependencies

```
"dependencies": {
  "node-fetch": "2.6.7"
},
"devDependencies": {
  "@types/node": "^17.0.21",
  "nodemon": "2.0.19"
},
```

3.　存檔

4.　照前述再上傳至 github

Environment 的值變成 node 了

5.　按 Fork 就可以正常開發了

## ▎ 7-6　Link Sandbox

我們可以將 github 上的專案，和我們開發中的 sandbox 藉由 Link Sandbox 功能，做一個同步的狀態。

1. 到 codesandbox 首頁選 New Sandbox

2. 選 Import Project

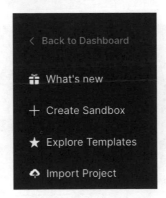

3. 回到 github 至 repository 頁，點 Code

4. 點 右邊 Code

5. 點  複製 HTTPS 的網址

6. 貼回剛才 Import from GitHub 的網址

7. 點 Import and Fork

8. 完成後，至

9. 點 Link Sandbox

10. 和 github 上的 repo 連結完成

## ▌7-7 加上 github repository 跟 npm 上的 package 的關聯性

通常我們會告知這個 github repository 跟 npm 上的 package 的關聯性，這樣其他開發者想替我們增加新 feature 或 debug 什麼時，才知道去那裡找 github repository。

## 7-7-1 改寫 package.json

1. 回到 github 至 repository 頁，點 Code

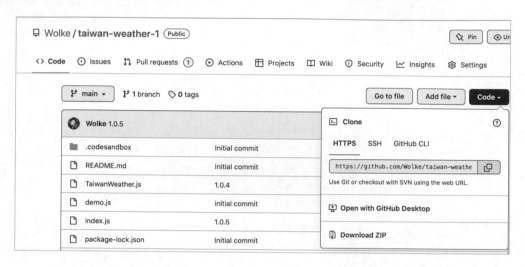

2. 點複製按紐 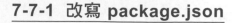 複製連結

3. 至 codesandbox 編輯 package.json

4. 將剛複製的連結，加上 `repository` 屬性

```
"repository": {
    "type": "git",
    "url": "https://github.com/xxx/taiwan-weather-xxx.git"
}
```

5. 版本號往上加，例如 1.0.1 變成 1.0.2 這樣才能更新

目前 package.json:

```
{
 "name": "taiwan-weather-xxx",
 "version": "1.0.2",
 "main": "TaiwanWeather.js",
 "license": "MIT",
 "dependencies": {
   "node-fetch": "2.6.7"
 },
 "devDependencies": {
   "@types/node": "^17.0.21",
   "nodemon": "2.0.19"
 },
 "repository": {
   "type": "git",
   "url": "https://github.com/xxx/taiwan-weather-xxx.git"
 }
}
```

## 7-7-2 NPM 再發佈一版

1. npm login 登入前述的帳號

```
npm login
```

```
sandbox@sse-sandbox-xns53m:/sandbox$ npm login
Username: wolke.web.app
Password:
Email: (this IS public)
Email: (this IS public) wolke.web.app@gmail.com
npm notice Please check your email for a one-time password (OTP
)
Enter one-time password from your authenticator app: 86507037
Logged in as wolke.web.app on https://registry.npmjs.org/.
```

2. npm publish 發布出去

```
npm publish
```

3. 到 npm 頁上看，多了 Repository 了

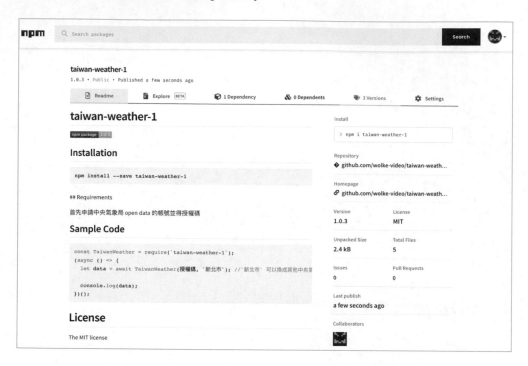

## 7-7-3 改寫 README.md

1. 回到 codesandbox 上，編輯 README.md

2. 加上

```
[![npm version](https://badge.fury.io/js/taiwan-weather-xxx.svg)]
(https://badge.fury.io/js/taiwan-weather-xxx)
```

目前 README.md:

```
# taiwan-weather-xxx

[![npm version](https://badge.fury.io/js/taiwan-weather-xxx.svg)](https://
badge.fury.io/js/taiwan-weather-xxx)

## Installation

'''bash
npm install --save taiwan-weather-xxx
'''

## Requirements

首先申請中央氣象局 open data 的帳號並得授權碼

## Sample Code

'''js
const TaiwanWeather = require('taiwan-weather-xxx');
(async () => {
 let data = await TaiwanWeather( 授權碼 , ' 新北市 '); //' 新北市 ' 可以換成其
他中央氣象局授受的城市名稱

 console.log(data);
})();
'''

# License

The MIT license
```

## 7-7-4 更新 github

1. 點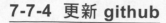

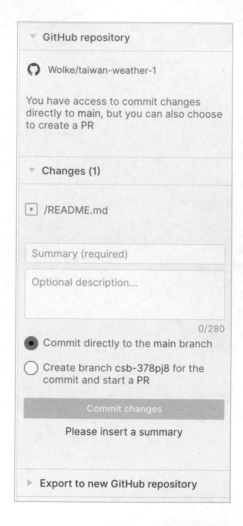

預設是 Commit directly to the main branch ：意思是送出至原本的 main 分支上。

2. Summary 必填，通常會是一個修改的小總結，而 description 則是選填，通常會完整說明，這裡我們可以在 summary 填 add npm badge 說明我們做了什麼的更動。

3.  點 Commits changes

4.  重新整理 github 的 repo 頁

5.  會發現多了一個 npm package 的圖片，上面註明目前的版本號，點這個圖片可以連回 npm package 的頁面。

6.  完成

# 7-8 協作開發

目前為止，我們的專案，不管是在 github 上，或是 npm 上都是只有我們一個開發者。

那要如何讓其他人如何透過 github 來協助開發呢？

## 7-8-1 透過 github 開 sandbox

為了了解如何讓其他人協作的整個運作方式，建議可以請至 github 再註冊一個新的帳號，並且用此新帳號登入 codesandbox。

1.  到 codesandbox

2.  點 create sandbox　Create Sandbox

3. 選 import

4. 將前面複製的連結填入 https://github.com/xxx/taiwan-weather-xxx.git

5. 點 Import and Fork

6. 好了後，點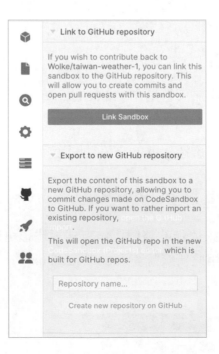

7. 點 Link Sandbox

## 7-8-2 發 PR pull request

當其他開發者開發完成之後，想要讓原本的擁有者獲得他的成果時，就可以發 PR 給對方了。

我們來演練整個流程。

1. 先開一個檔案叫做 demo.js

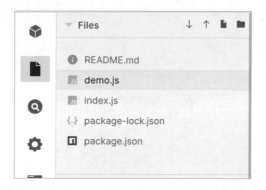

2. 可以看到 ⬚ 圖示亮燈，表示有檔案被修改了

3. 點 ⬚

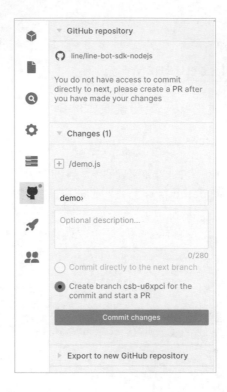

預設是 Create branch xxx for the commit and start a PR 和前面不一樣了，意思是新增一個新的 xxx 分支在你的 github 帳號上，並且發送一個 PR 至原本的 repo 顯示。

4. summary 填 demo

5. 點 Commits changes

6.  點 Open PR，會開啟到原本 repo 上

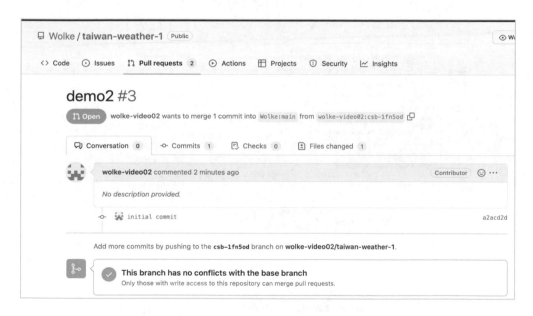

7.  可以看到剛剛發送的 PR 已經在原本 repo 的表列上了。

## 7-8-3 合併 PR pull request

1.　使用原本的帳號,開啟 PR 頁

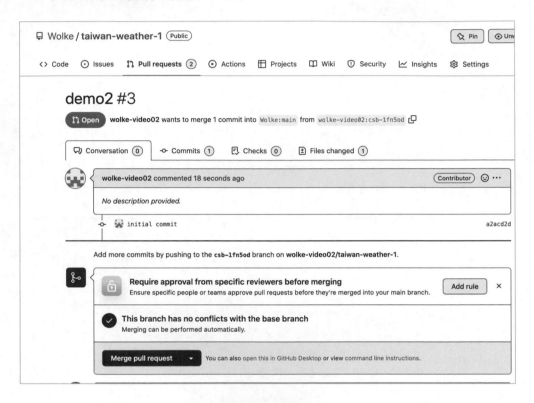

2.　會看到多了一個按鈕 `Merge pull request ▼`
3.　你可以選 Merge PR 去合併至目前的 repo ,也可以 Close PR 拒絕掉對方的 PR 請求
4.　我們先試著合併 `Merge pull request ▼`
5.　寫個合併原由

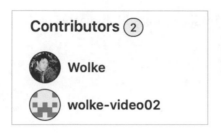

6.　合併完成後，回到首頁

7.　就會看到 demo.js 出現了，Contributors 也多了一個人

## 7-8-4　小結

這整個過程，就是一般團隊的程式協同開發作業一個流程體驗而已。

因為這裡是配合 codesandbox 及 github 的所提供的相關介面工具所演繹過程，所以是相當陽春簡化的。

通常還會搭配 git 來做版控等指令的操作。

但要講解版控的觀念等，恐怕不是一時三刻可以說的明白的，想再研究，可以再參考 git 及 github 的相關書籍。

## ▌ 7-9 Open Source 開源生態圈

開源：開放程式碼，本書著重在讓讀者了解開源生態圈，如果沒有這個生態體系的運行，程式相關的技術，也不會因為有著眾人的智慧結晶，而如此的一日千里。

例如本書著重在 node.js 的部份，node.js 在 2009 年寫成，並在 2010 年 NPM 套件管理問世，NPM 的問世使著程式設計師能方便地釋出分享自己的成果，以及使用其他程式設計師所撰寫的套件，而不需要再去重構一樣的功能，這加速了程式工程開發的速度。

許多套件，在撰寫過程裡，都會再引用到其他人所寫的套件，加快自己程式的開發工程，例如本所舉例的 express.js 也相依了許多套件，使 express 這個 framework 更為完整。

https://github.com/expressjs/express/blob/master/package.json

## 7-9-1 開源的好處

### ＊ 免費

筆者尤記得以前學習程式，光是 framework 授權費，就是一筆開支，但使用開源 framework ，例如 express ，你不需要先支付一毛錢， 就可以開始學習撰寫應用，甚至上架網站。

### ＊ 不用重新發明輪子

許多開發上的問題，早就被其他的開發者遇過，他們會撰寫成 package 發佈，而你只要引用過來，就能使用了。

### ＊ 學習程式撰寫

在初學程式語言的階段，參考大神的撰寫風格，絕對是使自己快速進步的方式。

## 7-9-2 開源的挑戰

聰明如你，應該也有發現，開源有很多問題。

## ＊ 免費

免費使用 framework，雖然使引用導入的開發者開發速度加快，但對於 framework 的擁有者來說，除非他們吸空氣就會飽，不然完全的免費都收不到錢，是要全家都跟著喝西北風嗎？

例如之前的新聞事件：

摘自：

https://www.ithome.com.tw/news/148822 ？ fbclid=IwAR3sy_X6t9Sn5BYLgLLgbHS29UviaF1OjwUEaS0QuK0ih5O3OceBlvVvguY

NPM (Node Package Manager) 線上套件庫兩個受歡迎的函式庫遭作者刻意破壞，使數千個使用這些函式庫的應用程式無法使用或出現亂碼。

兩個出問題的函式庫為 colors 及 faker。前者可為 node.js console 定義文字色彩，後者則可用於產生測試時的大量假數據。兩者下載次數各超過 2,000 萬及 280 萬，在 GitHub 上分別有 4,500 及 1,000 顆星。此外與之相依的專案各有 1.9 萬及 2,500 多個。這兩個函式庫都是由 Marak Squires（代號 Marak）的開發人員開發維護。

但本周使用兩個函式庫的應用程式或系統出現怪事。有一堆亂碼。

```
> cdk
LIBERTY LIBERTY LIBERTY
LIBERTY LIBERTY LIBERTY
LIBERTY LIBERTY LIBERTY
                  !          H|H|H|H|H          H
testing testing testing testing test ing t esting testing
...
```

原本媒體以為是駭客攻擊事件，但追查發現，「兇手」正是函式庫作者 Marak 本人。

雖然乍看之下很惡意，但 Marak 指出，和大多數人一樣，他也有家要養，有帳單要付。他決定要針對 faker 專案收費以支持未來開發。

破壞自己函式庫的兇手是作者本人，這樣的是事情只會層出不窮，情節輕微的就像是本文，嚴重的，誰知道會不會放個木馬，將公司內部機密文件通通搬出來呢。

## * 安全性

使用第三方套件加快開發速度，但通常開發不會去審視第三方套件的漏洞，也不知道第三方套件的漏洞，會不會造成什麼問題。

例如：

摘自：

https://www.secretchina.com/news/gb/2020/09/28/947459.html

據資安業者 "Check Point" 近期發布的研究報告指出，Instagram 照片分享平台目前每月在全球的活躍用戶近十億，人數十分龐大。而該應用程式所採取的第三方函式庫 Mozjpeg 被發現存在一項嚴重的安全漏洞，使得 Instagram 這個應用程式可能會遭到駭客利用。

IG 這麼大的公司，內部資安團隊肯定不會少，都中了第三方套件的虧了。

## * 新版更新了什麼？

如前述，雖然程式都是開源的，都是真的有人會每次都檢視開源內容嗎？

例如前端知名開源套件 ant design 在 2018 年聖誕節埋了一個彩蛋，但不知情的前端工程師們可慘了。

例如：

摘自：https://tech.sina.com.cn/csj/2018-12-26/doc-ihmutuee2734117.shtml

Antd 即 Ant Design，是阿裡螞蟻金服開源的一套企業級的 UI 設計語言和 React 實現，因提供了完整的類型定義檔、自帶提煉自企業級中後臺產品的交互語言和視覺風格、開箱即用的高品質 React 元件與全鏈路開發和設計工具體系等功能優勢而備受開發者青睞。

這次惹起爭議的就是 Antd 庫中暗藏的聖誕彩蛋——一個不曾提前告知、無法手動禁止、並且 " 潛伏 " 了 106 天之久的按鈕 " 驚喜 "，打得開發者們措手不及。更有甚者，因為很多政府專案、軍事專案、宗教專案等嚴肅場景中也一個不落地中了招，導致許多專案方遭遇客戶投訴，開發者們背鍋被裁。

一觸即發！一個 " 俏皮 " 的積雪效果按鈕引發開發者危機

那這個所謂的聖誕彩蛋究竟是指什麼？其實就是一個帶有積雪特效的按鈕，以及一個 "HoHoHo!" 的提示。

可以想見發生的當下，使用者、老闆及前端工程師們有多傻眼。

## 7-9-3 小結

我們習慣於相信 merge PR 的那位，會幫我們把關，但如果在營運的那位，都是無償經營了，發生這些事情，我們又何忍怪之？

## 7-9-4 開源相關的工具及平台

### ＊ Github

如今 github 就等同於開源的代名詞，而開發者的 github profile，就好像明星所經營的 ig profile 一樣，是大家認識你的第一印象。

時至今日，很多技術主管在應徵新人時，第一件事所參考的不是新人的學經歷，而是新人的 github profile，在今時今日 github 之於 IT 產業的重要性，以不言可喻。

例如筆者申請 GDE (Google Developer Expert) 時，也附上了我的 github 的連結，作為加分的條件，今時今日在 IT 產業打滾，沒有經營好自己的 github profile 實屬不智。

### ＊ COSCUP 開源人年會

https://coscup.org/

COSCUP 是由台灣開放原始碼社群聯合推動的年度研討會，會議內容是公開的，參與也是免費的，對於開源議題有相關興趣的朋友，記得每年關注，可以了解目前世界與台灣的各項趨勢。

## 7-9-5 開源練習

1. 將 game.js 開發成一個新套件，將其發佈至 NPM 上及開源上傳至 github 上。
2. 試自己找 github 上任一個開源庫 import 至 codesandbox 上，並修改內容後，發 PR 上去原開源庫試試

# 8

# LINE Bot 篇
# LINE Message
# API

## ▌ 8-1 前言

在 APP 開發式微之後，mobile 應用的開發主流，取而代的可以說是 LINE 聊天機器人，因為其優點一、免安裝，加好友就能進駐使用者手機；二、只要經過使用者授權，就可以獲得使用者資料，可以說是其優點二。

本章將會以目前所了解的知識作為基礎，藉由 LINE Message API 來實作建立規則 ruled base 的聊天機器人。

## ▌ 8-2 學習目標／演練成果

實作出氣象小幫手 LINE Bot 版，藉此了解整個 LINE Bot 的程式的撰寫

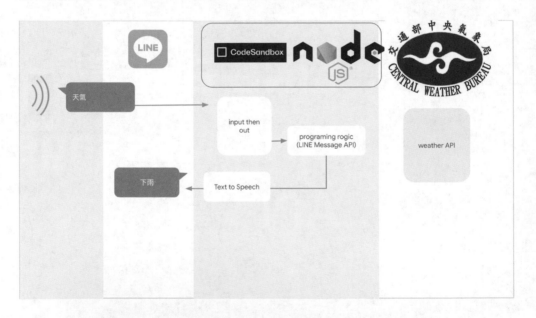

## ▌ 8-3 實做 天氣小幫手 LINE bot 版

### 8-3-1 申請 LINE Bot 開發者帳號

1.　首先 google LINE developer

2. 點擊 LINE Developers

3. 點 Login

4. 直接用你的 LINE 帳號登入

5.  如果是第一次註冊，應該會是空的。

6.  點 Create new provider

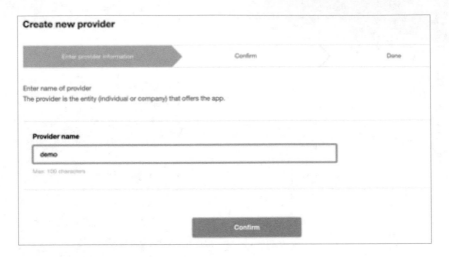

7.  Provider name 可以先填 demo

8.  點 Confirm 送出

9.  進來後 點 Create a new channel

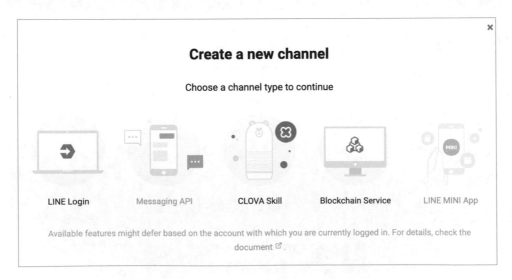

10. 目前 LINE 有多種服務可以開發，選 Messaging API 送出

11. 進來後，就是 LINE Bot 的基本資料的填寫

12. Company or owner's country or region ( 國家 ) : 可以選 Taiwan

13. Channel icon ( LINE Bot 的大頭貼 ) : 是選填的，可以上傳你喜歡的照片

14. Channel name ( LINE Bot 的名字 ) : 填想要取的名字

15. Channel description ( LINE Bot 的說明 ) : 填關於這個 LINE Bot 的相關說明

16. Category ( 主分類 )

17. Subcategory ( 次分類 )

18. Privacy policy URL 隱私權政策的網址 : 選填

19. Terms of use URL 使用條款 : 選填

20. I have read and agree to the LINE Official Account Terms of Use：必須打勾

21. I have read and agree to the LINE Official Account API Terms of Use：必須打勾

22. 填完後，點 Create **Create**

**Create a Messaging API channel with the following details?**

Channel name : demo
Official Account name : demo
Provider : demo

- If you proceed, an official account will be created with the same name as the messaging API channel above.
- You cannot change the channel provider after the channel is created. Make sure that the provider and official account owner are the same individual developer, company or organization.
- For the handling of LINE user information, please refer to User Data Policy ☐ .

| Cancel | OK |
| --- | --- |

23. 看一看，點 OK

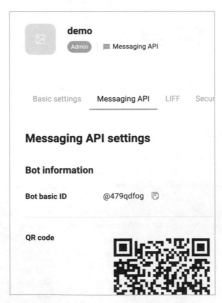

### 同意我們使用您的資訊

LINE Corporation（下稱「LINE」）為了完善本公司服務，需使用企業帳號（包括但不限於 LINE 官方帳號及其相關 API 產品;以下合稱「企業帳號」）之各類資訊。若欲繼續使用企業帳號，請確認並同意下列事項。

■ **我們將會蒐集與使用的資訊**
- 用戶傳送及接收的傳輸內容（包括訊息、網址資訊、影像、影片、貼圖及效果等）。
- 用戶傳送及接收所有內容的發送或撥話格式、次數、時間長度及接收發送對象等（下稱「格式等資訊」），以及透過網際協議通話技術（VoIP；網路電話及視訊通話）及其他功能所處理的內容格式等資訊。
- 企業帳號使用的 IP 位址、使用各項功能的時間、已接收內容是否已讀、網址的點選等（包括但不限於連結來源資訊）、服務使用紀錄（例如於 LINE 應用程式使用網路瀏覽器及使用時間的紀錄）及隱私權政策所述的其他資訊。

■ **我們蒐集與使用資訊並提供給第三方的目的**
上述資訊將被用於（i）避免未經授權之使用；（ii)提供、開發及改善本公司服務；以及（iii)傳送廣告。
此外，我們可能會將這些資訊分享給 LINE 關係企業或本公司的服務提供者及分包商。
如果授予此處同意的人不是企業帳號所有人所授權之人，請事先取得該被授權人的同意。如果 LINE 接獲被授權人通知表示其未曾授予同意，LINE 得中止該企業帳號的使用，且不為因此而生的任何情事負責。

同意

24. 看一看，點 同意

25. 新增 LINE Bot 完成

26. 到 Messaging API 頁，可以看到 Bot ID 以及 QR code，可以先加好友

27. 會看到 LINE 預設的訊息跳出

28. 輸入任何文字只會看到預設的訊息跳出

29. 接著到 Messaging API 頁，可以看到 Webhook settings 而 Webhook URL 還是空著的。

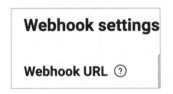

30. LINE Bot 就是依靠 webhook 來和我們的後端解決方案作溝通，下一步來用 codesandbox 建置 LINE Bot 伺服器

31. 但先將下面的 LINE Official Account features 裡的 Auto-reply messages, Greeting messages 都設為 Disable，就不會跳出預設訊息了。

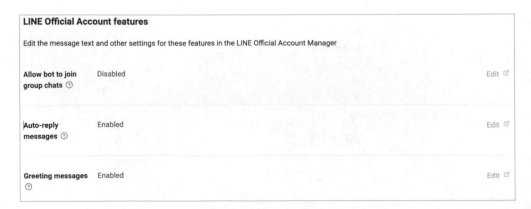

32. 點其中一個 Edit 會另開新頁至 LINE Offical Account Manager

33. 設定成如上就可以了

## 8-3-2 建置 LINE Bot 伺服器

1. 至 codesandbox 開 node http server，安裝完 nodemon 且改寫 package.
   json 裡的 start 為 nodemon index.js

```
"dependencies": {
},
"scripts": {
  "start": "nodemon index.js"
},
"devDependencies": {
  "nodemon": "2.0.19",
  "@types/node": "^17.0.21"
}
```

2. 點 restart sandbox 讓 package.json 裡的 scripts 的 start 的設定運行

3. 刪光 index.js 裡原有的 code

4. 安裝 express 套件並運行之

目前 index.js：

```
const express = require('express');
const app = express();
app.get('/', (req, res) => {
 res.end('hello');
});
app.listen(8080);
```

5. 試一下訪問 / 回傳 hello

6. LINE Bot 依靠 webhook 來做溝通，定義 webhook 路由

目前的 index.js：

```javascript
const express = require("express");
const app = express();
app.get('/', (req, res) => {
 res.end('hello');
});
app.post('/webhook', (req, res) => {
 console.log(req); // 將 LINE 傳來的內容印出來看一下
 res.end('ok');
});
app.listen(8080);
```

7. 將網址含 webhook 貼到 LINE Bot 的 webhook 設定頁

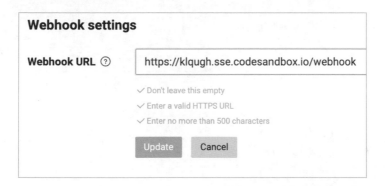

8. 點 Update

9. use webhook 打開

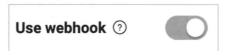

10. 跟 LINE Bot 隨便說些什麼

```
Terminal  36        Console  0        Problems  0
                      yarn start
route: Route {
  path: '/webhook',
  stack: [ [Layer] ],
  methods: { post: true }
},
[Symbol(kCapture)]: false,
[Symbol(RequestTimeout)]: undefined
}
```

11. 會發現傳來一堆內容，但沒發現我們輸入到 LINE Bot 的訊息，這是因為 LINE 有做一個加密的動作，我們必須要解密才能看到我們所傳送的訊息。

## 8-3-3 解密

1. 安裝 @line/bot-sdk 套件，並引用進 index.js

```
const line = require('@line/bot-sdk');
```

2.   LINE Bot 的訊息解密必須藉由 Access token 及 Secret

3.   到剛剛前面新增的 LINE Bot 的 Basic settings 頁，下方有一個 Channel secret

4.   複製 Channel Secret ； 按 Issue 可以換一組新的

5.   回到剛剛的 sandbox ，新增 secret 值

6.   再至 LINE Bot 的 Messaging API 頁 ，下方有一個 Channel access token 欄

7. 點擊 Issue 就會產生一組 token

8. 複製 Channel access token ；按 Reissue 可以換一組新的

9. 回到剛剛的 sandbox ，新增 token 值

10. 在 index.js 裡宣告 config 值

```
const config = {
  channelAccessToken: process.env.token,
  channelSecret: process.env.secret
};
```

11. 可以試著加一行 console.log 看值有沒有印出來

```
console.log(config);
```

12. 在剛才定義的 webhook 路由裡，中間加上 line.middleware(config) 做為
中介軟體來做轉譯的動作

```
app.post('/webhook',line.middleware(config) , (req, res) => {
    console.log(req.body);
})
```

13. 跟 LINE Bot 隨便說些什麼，可以看到 req.body 有值出來了，這是因為 req 第一次傳進來後，會先被 line.middleware 做轉譯，產生出 req.body，再吐出來給 (req, res) => {}

14. 其中的 req.body.events 就是 LINE 傳過來的訊息表列。

15. events 為一個陣列，裡面就是使用者或 LINE 平台傳來的訊息。

16. 印出 req.body.events[0].message

```
console.log(req.body.events[0].message);
```

17. 跟 LINE Bot 說你好，可以在 terminal 看到訊息

## 8-3-4 回傳訊息

在 LINE Bot 中將傳訊息給使用這個行為，分為兩種

push：直接傳給使用者

reply：使用者得先傳訊息過來，方能回覆使用者

push 是被 LINE 所管制的，reply 則是可以無限使用，這裡我們會演繹兩種。

### ✳ reply API

1. 先宣告 line 的 Client 物件，來跟 LINE 平台做為通溝的橋梁。

```
const client = new line.Client(config);
```

2. 我們剛剛在印出 console.log(req.body); 有看到 events 裡有一個 replyToken 的值，我們就是要藉由 replyToken 來呼叫 reply api。

```
[nodemon] starting `node index.js`
{
  destination: 'Ue1af376c96854712751e6ca7c32bfaba',
  events: [
    {
      type: 'message',
      message: [Object],
      webhookEventId: '01G8A970HZE0N1RH1QXMS8NWT3',
      deliveryContext: [Object],
      timestamp: 1658202587284,
      source: [Object],
      replyToken: '6c4d370482374440b68412d4a4e5c013',
      mode: 'active'
    }
  ]
}
```

3. 直接將使用者傳來的訊息發送回去

4. 先宣告 msg

```
let msg = {
  type: "text",
  text: req.body.events[0].message.text // 使用者傳來的訊息
};
```

5. 呼叫 client.replyMessage 將 msg 發送回去

```
client.replyMessage(req.body.events[0].replyToken, msg);
```

6. 完成

目前的 index.js：

```
const express = require("express");
const line = require("@line/bot-sdk");
const app = express();
const config = {
 channelAccessToken: process.env.token,
 channelSecret: process.env.secret
};
const client = new line.Client(config);

app.get('/', (req, res) => {
 res.end('hello');
});
app.post("/webhook", line.middleware(config), (req, res) => {
 let msg = {
   type: "text",
   text: req.body.events[0].message.text // 使用者傳來的訊息
 };
 client.replyMessage(req.body.events[0].replyToken, msg);
});
app.listen(8080);
```

7. 試一試

## ✳ push API

想要直接發送訊息給使用者，必須將不同使用者的 user id 給記起來，要注意的是，就算是同一個使用者在不同的 LINE Bot 中，都會拿到不同的 user id 的值。

這是因為 LINE 平台會虛擬出不同的值給不同的 LINE Bot，所以 user id 無法通用喔。

1. user id 在 events 裡面的 source 就有

```
[nodemon] starting `node index.js`
{
  destination: 'Ue1af376c96854712751e6ca7c32bfaba',
  events: [
    {
      type: 'message',
      message: [Object],
      webhookEventId: '01G8A970HZE0N1RH1QXMS8NWT3',
      deliveryContext: [Object],
      timestamp: 1658202587284,
      source: [Object],
      replyToken: '6c4d370482374440b68412d4a4e5c013',
      mode: 'active'
    }
  ]
}
```

2.　印出來

```
console.log(req.body.events[0].source);
```

3.　記住這個 userId 就可以呼叫 push 直接發送訊息給他了

```
client.pushMessage("Ubfcbff3466d011fa291…", {
  type: "text",
  text: ' 你好 Hello '
});
```

4.　完成

目前的 index.js:

```javascript
const express = require("express");
const line = require("@line/bot-sdk");
const app = express();
const config = {
 channelAccessToken: process.env.token,
 channelSecret: process.env.secret
};
const client = new line.Client(config);
client.pushMessage("Ubfcbff3466d011fa291050bb5cd73c0c", {
 type: "text",
 text: ' 你好 Hello '
});

app.get('/', (req, res) => {
 res.end('hello');
});

app.post("/webhook", line.middleware(config), (req, res) => {
 console.log(req.body.events[0].source);
 let msg = {
   type: "text",
   text: req.body.events[0].message.text // 使用者傳來的訊息
 };
 let d = client.replyMessage(req.body.events[0].replyToken, msg);
 res.json(d);
});
app.listen(8080);
```

## 8-3-5 引用 天氣小幫手套件

### ＊ 自動發送給使用者

1. 安裝 taiwan-weather-xxx 套件，並引用進 index.js

```javascript
const line = require(' taiwan-weather-xxx');
```

2. 新增氣象局授權碼至 Secret key

3. 呼叫並發送給使用者

```
(async () => {
 let data = await TaiwanWeather(process.env.key, '臺南市'); // 引用 KEY，
並指定查詢城市
 let { city, degree, weather, feeling, chance } = data;
 client.pushMessage("Ubfcbff3466d011fa291050bb5cd73c0c", {
   type: "text",
   text: '${city} 今天的天氣為 ${weather} 感覺 ${feeling} 最高溫為 ${degree.
max} 度 最低溫為 ${degree.min} 度 降雨機率為 ${chance} % '
 });
})();
```

4. 重啟就會收到訊息

目前的 index.js:

```javascript
const express = require("express");
const line = require("@line/bot-sdk");
const TaiwanWeather = require('taiwan-weather-1');

const app = express();
const config = {
 channelAccessToken: process.env.token,
 channelSecret: process.env.secret
};

const client = new line.Client(config);

(async () => {
 let data = await TaiwanWeather(process.env.key, '臺南市'); // 引用 KEY，
並指定查詢城市
 let { city, degree, weather, feeling, chance } = data;
 client.pushMessage("Ubfcbff3466d011fa291050bb5cd73c0c", {
   type: "text",
   text: '${city} 今天的天氣為 ${weather} 感覺 ${feeling} 最高溫為 ${degree.
max} 度 最低溫為 ${degree.min} 度 降雨機率為 ${chance} % '
 });
})();

app.get('/', (req, res) => {
 res.end('hello');
});
app.post("/webhook", line.middleware(config), (req, res) => {
 console.log(req.body.events[0].source);
 let msg = {
   type: "text",
   text: req.body.events[0].message.text // 使用者傳來的訊息
 };
 let d = client.replyMessage(req.body.events[0].replyToken, msg);
 res.json(d);
});
app.listen(8080);
```

＊ **依據參數發送**

當然也可以讓使用者輸入想查詢的城市來發送天氣狀況

1. 在 wcbhook 路由拿使用者輸入的城市

```
let { text } = req.body.events[0].message;
```

2. 查找資料

```
let data = await TaiwanWeather(process.env.key, text); // 引用 KEY，並指定
查詢城市
let { city, degree, weather, feeling, chance } = data;
```

3. 回傳使用者

```
let msg = {
   type: "text",
   text: '${city} 今天的天氣為 ${weather} 感覺 ${feeling} 最高溫為 ${degree.
max} 度 最低溫為 ${degree.min} 度 降雨機率為 ${chance} % '
 };

 let d = client.replyMessage(req.body.events[0].replyToken, msg);
```

4. 試一試

## ＊ 想一想

LINE Bot 查找資料的部份，當使用者輸入非預期的名稱時，就會導致程式失效，嚴謹一點可以做個使用者文字的對應

5. 確認使用者輸入的是否有這個城市，沒有的話，就回傳沒有

```
let citys = [
    '宜蘭縣',
    '花蓮縣',
    '臺東縣',
    '澎湖縣',
    '金門縣',
    '連江縣',
    '臺北市',
    '新北市',
    '桃園市',
    '臺中市',
    '臺南市',
    '高雄市',
    '基隆市',
    '新竹縣',
    '新竹市',
    '苗栗縣',
    '彰化縣',
    '南投縣',
    '雲林縣',
    '嘉義縣',
    '嘉義市',
    '屏東縣'
];
if (!citys.includes(text)) {
    let d = client.replyMessage(req.body.events[0].replyToken, {
        type: "text",
        text: '沒有 ${text} 這個地方的天氣資料喔'
    });
    res.json(d);
    return;
}
```

6. 試一試

## ▌ 8-4　小結

前面使用判斷式來判斷使用者所輸入的資料，就是所謂的 ruled base 聊天機器人。

想一想，氣象局認列的臺北市不是我們習慣用的台北市，是不是得在使用者輸入時，在程式裡做轉換，避免出錯呢？

```
if(text === 台北市 ){
    text = 臺北市
}
```

此外有些約定俗成的名稱是不是也要做轉換避免出錯

```
if(text === 台北 ){
    text = 臺北市
}
```

而且，在功能相對陽春時，還可以這麼做，但當功能越來越多時，還能這麼做嗎？ 例如這個聊天機器人可以查天氣，也可以玩遊戲時呢？

是不是判斷式會越寫越多，也越來越雜呢？

例如：

```
if( 天氣 ){
  if( 今天 ){
    ...
  }else if( 明天 ){
    ...
  }
}else if( 遊戲 ){
    if( 剪刀石頭布 )}{
      ...
    }
}else if( 交友 ){
    ...
}
```

可以想像如此處理下去，程式碼會有多少判斷式。

又或是使用者輸入了：今天氣死了，但他不是要問天氣，卻因為有天氣兩個字，而判斷錯誤；諸如此類的問題發生，這個時候，我們就要導入 AI 來協助我們，下一章我們將會來介紹 NLU 語言人工智能模組 Google Dialogflow

# 8-5 常用功能解釋

## 8-5-1 webhook

webhook 的主要功能就是給 LINE 平台傳遞 event 給我們用的。

在很多其他應用的開發，只要需要讓平台傳遞訊息給後端的，都會藉由 webhook 的機制來做訊息傳遞動作。

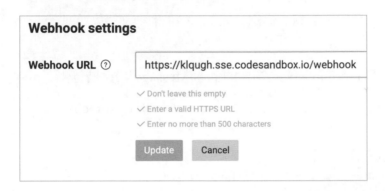

但是訊息傳遞的設計，自然是要加密的，當然你以後有機會設計 webhook，想要將其設計程成明碼也可以，只是難免要擔心日後會不會被有心人士給利用。

而 LINE 的 webhook 的加密方式也改過幾個版本了，猶記得最早期是在 req. rawBody 裡做加密的，在官方的 source code 裡還看的到這個部份。

https://github.com/line/line-bot-sdk-nodejs/blob/next/lib/middleware.ts

```
if (isValidBody((req as any).rawBody)) {...}
```

但目前的都改成包封包了。

這裡也可以看到使用套件的好處，雖然解密方式改變了，可變動只是 middleware.ts 的程式，我們的程式碼依舊呼叫 line.middleware 做解密的動作，原本的程式碼一行程式都不用改。

```
app.post('/webhook',line.middleware(config) , (req, res) => {
...
}
```

## 8-5-2  client

```
const client = new line.Client(config);
```

client 和 LINE 平台溝通使用，這個 client 的設制機制，也是很常在第三方平台的服務套件上看到，藉由生成 client 物件，將平台的服務，通通包裹在程式裡面。

例如前面的 LINE 的 client 物件的 pushMessage 其實是包裹了 reply API

https://github.com/line/line-bot-sdk-nodejs/blob/next/lib/client.ts

```
public replyMessage(
    replyToken: string,
    messages: Types.Message | Types.Message[],
    notificationDisabled: boolean = false,
  ): Promise<Types.MessageAPIResponseBase> {
    return this.http.post('${MESSAGING_API_PREFIX}/message/reply', { //
這裡帶著 messages , replyToken ZY 參數對 LINE 平台訪問了 reply 的網址
        messages: toArray(messages),
        replyToken,
        notificationDisabled,
    });
  }
```

另外，可以看到我們只用了 pushMessage 和 replyMessage 兩個函式，但除此之外，還有很多函式也是被包在裡面。

其他的函式，可以參考：

https://line.github.io/line-bot-sdk-nodejs/api-reference/client.html#methods

### 8-5-3 中介軟體 middleware

是一類提供系統軟體和應用軟體之間連接、便於軟體各部件之間的溝通的軟體。

express 本身就是路由與中介軟體的 Web 架構。

也因為 express 有中介軟體這樣的設計，因此 LINE 的官方套件，藉由設計出 LINE.middleware 做為中介軟體來做轉譯由 LINE 平台上所傳送過來的訊息。

```
// globally
app.use(middleware(config))

// or directly with handler
app.post('/webhook', middleware(config), (req, res) => {
  req.body.events // webhook event objects
  req.body.destination // user ID of the bot (optional)
  ...
})
```

### 8-5-4 push & reply api

LINE 將使用者傳訊的行為分為 push 跟 reply 其實主要跟 LINE 的商業策略有關，使用 reply 函式所發送訊息是幾乎沒有限制的；但使用 push 則是有限制。

可以參考 https://tw.linebiz.com/column/budget-auto-count/

在 Messaging API 提供 Push 與 Reply 兩種形式的訊息模式 API，一個需要付錢一個則不用，但兩個都需要非常注意：

● 「Reply API（自動回覆訊息）」：

指 LINE 機器人針對 LINE 用戶傳來的訊息進行自動回覆的 API，屬於「不列入計價的訊息」。

● 「Push API（主動推播訊息）」：

指 LINE 機器人可在任何時間點對用戶主動傳送訊息的 API，是要「列入訊息計價」，所以商家在使用由 LINE 機器人主動對好友發訊息的 Push API 時，要非常注意後台數據，以免不小心就超過訊息成本預算。

## 價格彈性

依照發送量計價，靈活配比預算

| | 低用量 | 中用量 | 高用量 |
|---|---|---|---|
| 固定月費 | 免費 | 800元 | 4,000元 |
| 免費訊息則數 | 500則 | 4,000則 | 25,000則 |
| 加購訊息費用 | 不可 | 0.2元 | 0.15元~ |

※未稅價格

## 8-5-5 Events

LINE 主要會從 webhook 傳送 event。

event 除了前面範例的 type 為 messsage 的 event 之外，還有被加好友或是被不加好友的 follow 和 unfollow 以及群組 group 用的 join 及 leave 等等。

關於 event 可以參考

https://developers.line.biz/en/reference/messaging-api/#webhook-event-objects

這些通通都會帶 replytoken 及 source 過來。

例如：

```
[
  {
    type: 'join',
    webhookEventId: '01G8FVEHR6C87KFYZ50DNXEN0N',
    deliveryContext: { isRedelivery: false },
    timestamp: 1658389481055,
    source: { type: 'group', groupId: 'Cf82a7fdc622f2d857fb9a98
b21cd4ac4' },
    replyToken: 'a136904745c740449346160a9d6371a6',
    mode: 'active'
  }
]
```

所以我們就可以利用 replytoken 去發訊息

```
client.replyMessage(req.body.events[0].replyToken, {
  type: "text",
  text: ' 誰拉我進群 '
});
```

## 8-5-6 Message objects

除了範例中發送的 text message 之外， LINE 還有 sticker , image , video , audio 等

關於 message 的種類可以參考

https://developers.line.biz/en/reference/messaging-api/#message-objects

我們可以發送除了文字之外的其他訊息，需要注意的是，需將 message 包成陣列，並且一次不能超過 5 則。

例如：

```
client.replyMessage(req.body.events[0].replyToken, [
  {
    type: "text",
    text: ' 你好 !'
  },
  {
    type: "sticker",
    packageId: "446",
    stickerId: "1988"
  },
  {
    type: "audio",
    originalContentUrl: "https://example.com/original.m4a",
    duration: 60000
  },
  {
    type: "location",
    title: "my location",
    address: "1-6-1 Yotsuya, Shinjuku-ku, Tokyo, 160-0004, Japan",
    latitude: 35.687574,
    longitude: 139.72922
  },
```

```
{
  type: "flex",
  altText: "this is a flex message",
  contents: {
    type: "bubble",
    body: {
      type: "box",
      layout: "vertical",
      contents: [
        {
          type: "text",
          text: "hello"
        },
        {
          type: "text",
          text: "world"
        }
      ]
    }
  }
});
```

## 8-5-7  Rich menu

rich menu 就是當我們和官方帳號也就是 LINE 對話時，在手機版的下方有的設計會對話框。

例如 疾管家 下方的欄位 就是 rich menu。

目前只支援手機版，也無法在群組或群聊中使用。

關於 rich menu 可以參考

https://developers.line.biz/en/docs/messaging-api/using-rich-menus/

1. 準備好一張做為 rich menu 的底圖

2. 定義好 richmenu 的格式

```
let richmenu = {
 size: { // 大小
   width: 2500,
   height: 1686
 },
 selected: false,
 name: "Nice richmenu",// 標題，不會顯示在手機上，僅管理方便
 chatBarText: "Tap here",// 說明，不會顯示在手機上，僅管理方便
 areas: [// 定義可作用範圍，可以定義多組
   {
     bounds: {// 範圍左上角
       x: 0,
       y: 0,
       width: 2500,
       height: 1686
     },
     action: {// 被按到時的行為，這是定義為傳送訊息，action 還有很多種可以定義，
詳 action 節
       type: "message",
       label: "hello",
       text: "hello"
     }
   }
 ]
};
```

3. 然後拿 richMenuId

```
let richMenuId = await client.createRichMenu(richmenu)
```

4. 引用內建模組 fs

```
const fs = require('fs');
```

5. 將圖片讀進來，並設定為 richMenuId 的底圖

```
let r = await client.setRichMenuImage(
   richMenuId,
   fs.createReadStream('./rm.png')
);
```

6. 設定為預設的 rich menu

```
client.setDefaultRichMenu(richMenuId);
```

7. 完成

## 8-5-8　Quick Reply

Quick Reply 就是快速回應，目前只支援在手機端，方便使用者快速輸入我們預期的選項。

只需要加在回傳的訊息格式裡，就可以在手機裡顯示出效果。

關於 Quick Reply 可以參考

https://developers.line.biz/en/docs/messaging-api/using-quick-reply/

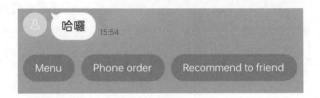

```
{
    "messages": [
        {
            "type": "text",
            "text": "Have you decided on your order ? ",
            "quickReply": {
                "items": [
                    {
                        "type": "action",
                        "action": {
                            "type": "uri",
                            "label": "Menu",
                            "uri": "https://example.com/menu"
                        }
                    },
                    {
                        "type": "action",
                        "action": {
                            "type": "uri",
                            "label": "Phone order",
                            "uri": "tel:09001234567"
                        }
                    },
                    {
```

```
                            "type" : "action" ,
                            "action" : {
                                "type" : "uri" ,
                                "label" : "Recommend to friend" ,
                                "uri" : "https://line.me/R/nv/recommendOA/@
linedevelopers"
                            }
                        }
                    ]
                }
            }
        ]
    }
```

## 8-5-9 Flex Message

主要用來發送 Card 的格式用的，可以藉由 CSS 語法來自訂想要的格式，主要是用來取代最開始的 Template Message 。

Template Message 都是固定格式的：

Flex message 則可以依照需求去做調整 CSS 語法，去達到想要的效果：

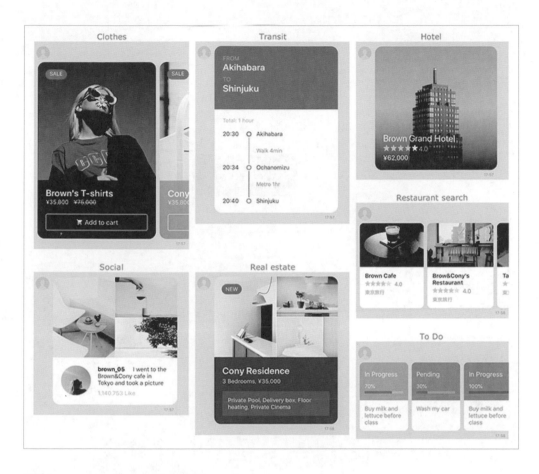

關於 Flex message 可以參考

https://developers.line.biz/en/docs/messaging-api/using-flex-messages/

這是一個簡單的 flex message

```
{
 type: "flex",
 altText: "This is a Flex Message",
 contents: {
   type: "bubble",
   body: {
     type: "box",
     layout: "horizontal",
     contents: [
       {
         type: "text",
         text: "Hello,"
       },
       {
         type: "button",
         action: {
           type: "uri",
           label: "Tap me",
           uri: "https://example.com"
         },
         style: "primary",
         color: "#0000ff"
       }
     ]
   }
 }
}
```

## 8-5-10 Actions

你可以看到 message 裡的回覆方式，都是使用 action 來做回覆

以前 LINE Message API 剛 release 時， action 的格式並不多，但目前已經較為完善了。

關於 actions 可以參考

https://developers.line.biz/en/docs/messaging-api/actions/

例如上述的範例，如果將 action 的內容改成 message ，就會變成按了就會回傳訊息 yes 出來。

```
{
        type: "button",
        action: {
          type: "message",
          label: "Yes",
          text: "Yes"
        },
}
```

## 8-5-1 Groups Chat

Groups Chat 群聊的 API 是 LINE Bot 一個特色，我們可以將 LINE Bot 置於群裡，去接收大家所發送的訊息，並發出相對應的反應；當然你也可以發送 push 到群裡，只是 push 的記算方式，並不是發送一次的這個行為，而是以接受到這個訊息的人數來做計算，想像一個 500 人群，發送一則訊息幾次，可能一下就到收費門檻了

所以如果寫 Groups Chat Bot 的話，盡量以 reply 來做發訊，會是較為明智的選擇。

關於 Groups Chat 可以參考

https://developers.line.biz/en/docs/messaging-api/group-chats/

這是接收到的 group 的 event

```
[
  {
    type: 'message',
    message: { type: 'text', id: '16485733349415', text: 'Hi' },
    webhookEventId: '01G8QNXRAP3YBGHEF04PJGG5NH',
    deliveryContext: { isRedelivery: false },
    timestamp: 1658652123106,
    source: {
      type: 'group',
      groupId: 'Cd611fbff6ec7b996ea5ac122d42e832a', // 群組的 ID
      userId: 'Ud15218793f171b2d41fb717b0035743f' // 發送者的 ID
    },
    replyToken: 'd76309bcdb6f476b9cacc725b70a2a7b',
    mode: 'active'
  }
]
```

你可以對群組 ID 發送 push，也可以依照不同的使用者 ID 去做不同的行為；但如果該群組內發送該訊息的使用者，尚未加你為好友，是無法去呼叫 getProfile 拿到該使用者的資料的喔。

## ▌ 8-6 附帶一提：程式工程師開發新內容的起手式

當我們新增 channel 之後，再來就是要開始開發了，點進官網看文件，就會
發現資訊爆炸，那怎麼辦呢？

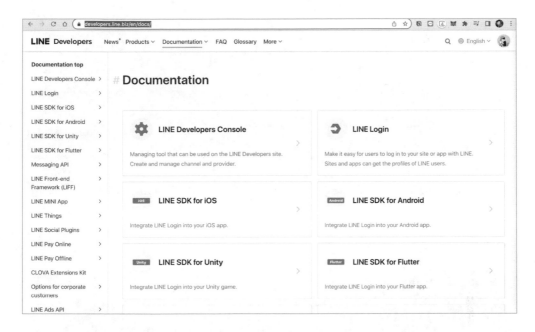

在這個應用該怎麼開發，一點概念都沒有的情況下，與其先看文件 ，被淹
沒在資訊海裡，不如先看 git repo 有什麼。

### 8-6-1 尋找 LINE Bot 的開發資源

在這個時代裡，幾乎所有的軟體公司都有 github 的官方帳號，我們可以直
接去 github 搜尋 LINE Bot 就可以找到 LINE 的官方帳號。

https://github.com/line

可以看到有各種程式語言的 sdk ，從名稱之中，我們可以了解 line-bot-sdk-nodejs 是 node.js 版本的 line bot sdk 。

點進去

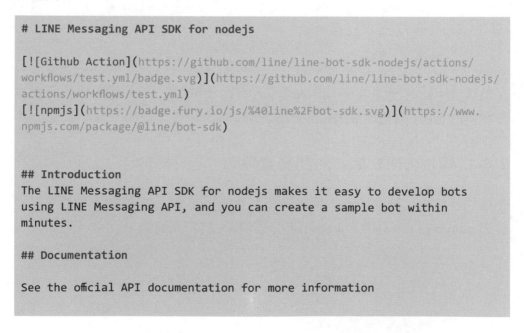

```
# LINE Messaging API SDK for nodejs

[![Github Action](https://github.com/line/line-bot-sdk-nodejs/actions/
workflows/test.yml/badge.svg)](https://github.com/line/line-bot-sdk-nodejs/
actions/workflows/test.yml)
[![npmjs](https://badge.fury.io/js/%40line%2Fbot-sdk.svg)](https://www.
npmjs.com/package/@line/bot-sdk)


## Introduction
The LINE Messaging API SDK for nodejs makes it easy to develop bots
using LINE Messaging API, and you can create a sample bot within
minutes.

## Documentation

See the official API documentation for more information
```

```
- English: https://developers.line.biz/en/docs/messaging-api/overview/
- Japanese: https://developers.line.biz/ja/docs/messaging-api/overview/

line-bot-sdk-nodejs documentation: https://line.github.io/line-bot-sdk-
nodejs/#getting-started
```

可以發現 README.md 裡，藏著一個 getting-started 教你從那頭開頭，通常就從這裡開始參考學習就沒錯了。

參　考 https://line.github.io/line-bot-sdk-nodejs/getting-started/basic-usage.html#configuration

# 9

## 機器學習篇
## Dialogflow

## ▍9-1 前言

如前一章 rule base 的 chat bot 雖然在功能上是可行的，但是對於使用者體驗來說是糟糕的，變成你得設計許多元件來讓使用者體驗較好，以至於後來許多的 chat bot messenger 都將前端元件給拉進來，例如 LINE 的 LIFF，微信的小程序這一類的應用。

但其實我們是可以將 chat bot 加上大腦，讓其辨識出我們的意圖。

## ▍9-2 學習目標／演練成果

dialogflow 的概念

dialogflow API 的使用

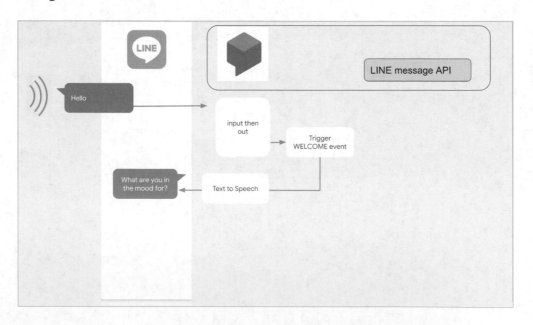

# 9-3 dialogflow 基本介面操作

## 9-3-1 使用 dialogflow

1. google dialogflow

2. ACCEPT 同意

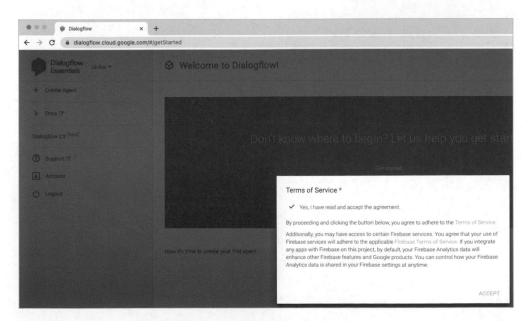

3. 建立 agent

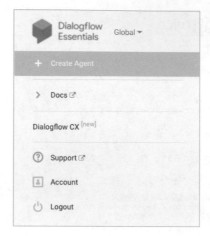

4. 命名 agent

5. 選語系 DEFAULT LANGUAGE 為 Chinese(Tradtional) zh-tw

6. 點 create

預設的 agent 有產生 2 個 intents，Default Welcome Intent 跟 Default Fallback Intent，intents 的意思翻成中文就是 意圖。

7. 進來後 點 welcome intent

8. 移到 Response 的部份，點刪除

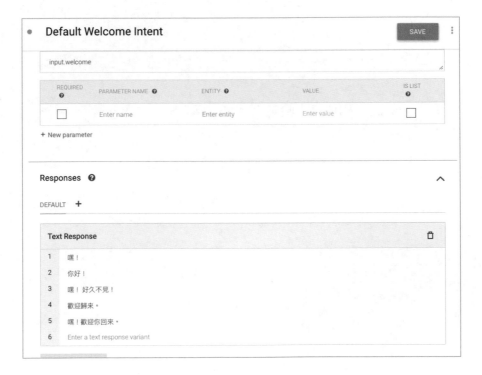

9. 新增 Text Response

10. 改成 呷飽沒

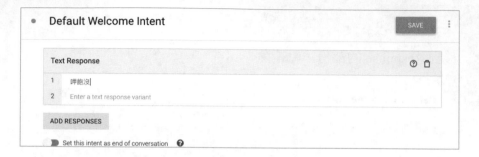

11. 在右上角的測試欄位打上哈囉

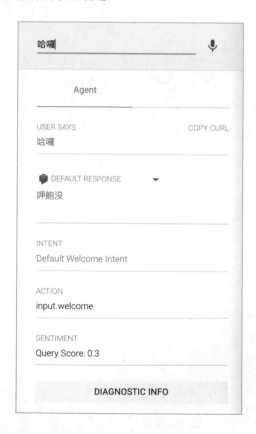

USER SAYS

DEFAULT RESPONSE

## 9-3-2 dialogflow 串接 LINE Bot

1. 在右邊選項裡點 Integrations

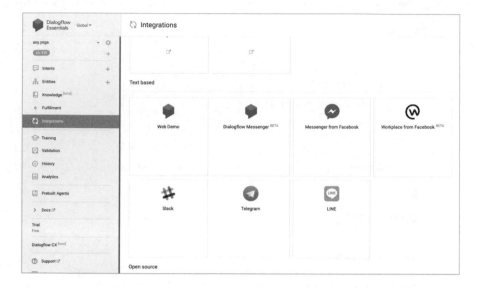

2. Integrations 頁裡，找一個 LINE 的選項

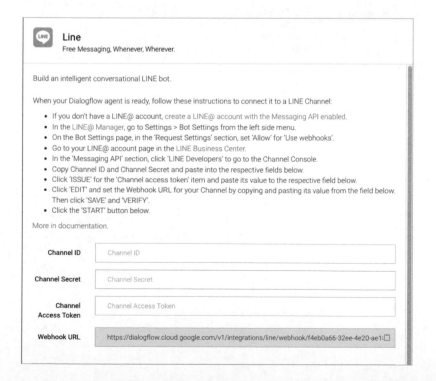

3. 點下去後，需要 Channel ID 跟 Channel Secret 跟 Channel Access Token

4. 註冊 LINE developer 帳號 ( 如前一章節所示 )

5. 我們新增一個 channel 後，將 Channel ID 跟 Channel Secret 跟 Channel Access Token 複製

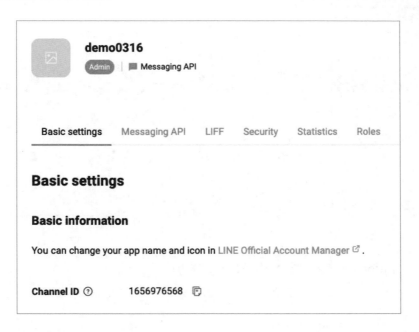

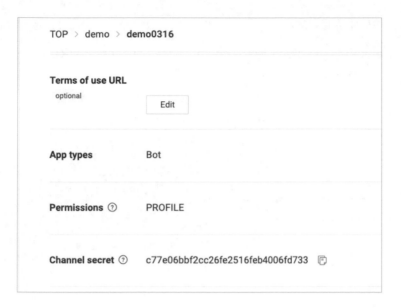

6. 將 Channel ID 跟 Channel Secret 跟 Channel Access Token 貼到 LINE 的對話框裡

7. 按 START

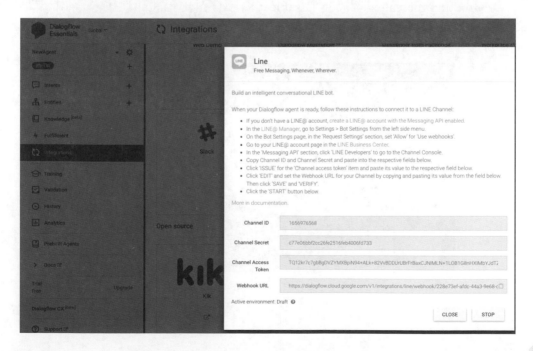

8. 複製 Webhook URL

9. 將 Webhook URL 貼回 LINE channel 裡的 Webhook URL 裡

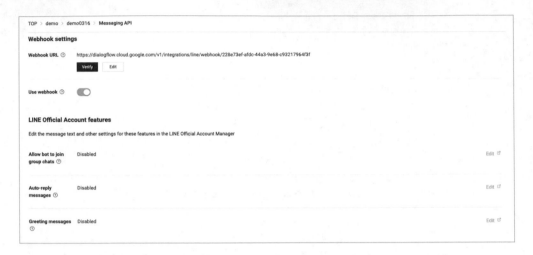

10. 將 Use webhook 開啟

11. 將下面的 LINE Official Account features 裡的 Auto-reply messages，Greeting messages 都設為 Disable，就不會跳出預設訊息了。

12. 完成 試一試

## 9-4 Intents 意圖

預設：

- Default Fallback Intent：Fallback 就是錯誤處理，所有認不出來的意圖，都會觸發到這裡。

- Default Welcome Intent ：Welcome 的意思就是打招呼的意思，可以在 Training phrases 裡看到已經預先訓練了許多語句。

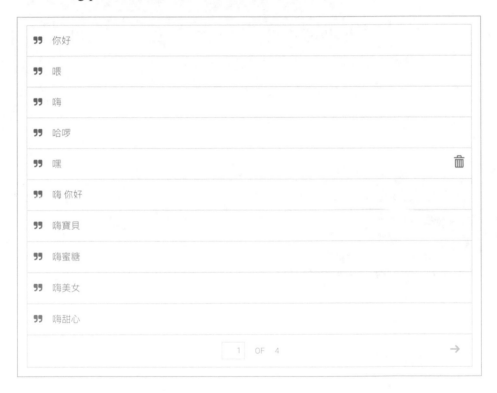

| 99 | 你好 | |
| --- | --- | --- |
| 99 | 喂 | |
| 99 | 嗨 | |
| 99 | 哈囉 | |
| 99 | 嘿 | 🗑 |
| 99 | 嗨 你好 | |
| 99 | 嗨寶貝 | |
| 99 | 嗨蜜糖 | |
| 99 | 嗨美女 | |
| 99 | 嗨甜心 | |

1 OF 4 →

前面的範例觸發了預設的 Default Welcome Intent ，這裡我們要新增一個 weather 的意圖，當使用者詢問關於天氣的事情時，都會觸發到這裡。

1. 點 Intents
2. 點 Create Intent
3. Intent name 取為 weather

4. Training phrases 裡盡量輸入問天氣相關的問題

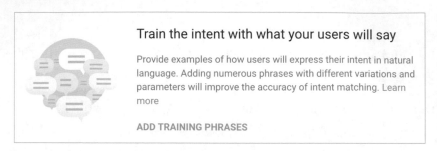

5. 點 ADD TRAINING PHRASES

6. 輸出訓練語句

7. 到 Response 輸入回覆的語句

8. 這裡先設定回覆 天氣 ⋯ 方便測試

9. 按 SAVE  SAVE

10. 測試一下

11. 試一些沒有輸入的語句試試有無 AI

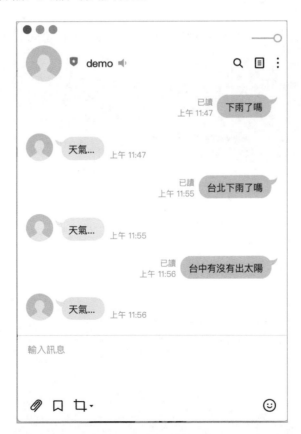

## 9-5 Entities

在串接天氣 API 之前，我們希望可以辨別出問的是那一個城市

1. 點 Entities

2. 點 CREATE ENTITY

3. Entity Name 命名 city

4. 輸入城市的值， reference value 的部份要對應氣象局的值。

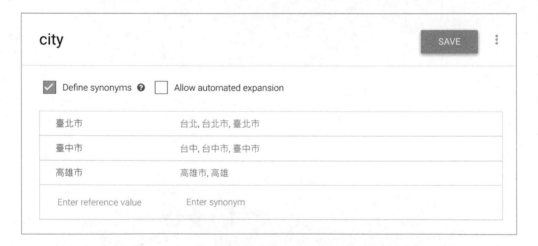

5. 點 SAVE

6. 回到 Intents 點 weather

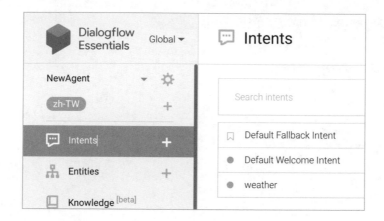

7.　輸入 台北天氣

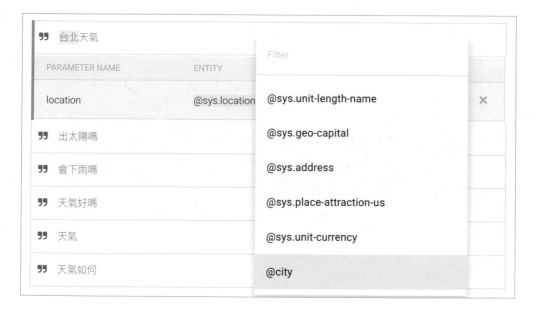

8.　點黃色的部份，選 @city

| PARAMETER NAME | ENTITY | RESOLVED VALUE | |
|---|---|---|---|
| city | @city | 台北 | × |

9.　到 Responses 的部份，新增 $city 的天氣

10.　點 SAVE

11.　試一試

# 9-6  call dialogflow API

這裡我們將會實作如何在 codesandbox 裡呼叫 dialogflow API 來協助我們判斷使用者所輸入的 Intent 和 entites

## 9-6-1  codesandbox 環境建置

1.  至 codesandbox 開 node http server，安裝完 nodemon 且改寫 package.json 裡的 start 為 nodemon index.js

2.  新增 intentDetect.js 作為意圖偵測之用

3.  宣告 intentDetect 的函式作為意圖偵測並 export 出去

```
const intentDetect = (text) => {
 console.log(text);
};
module.exports = intentDetect;
```

4.  為了方便開發，下面先宣告一個會執行的函式並呼叫 intentDetect 函式

```
const intentDetect = async (text) => {
 console.log(text);
};
module.exports = intentDetect;
/* */
(() => {
 intentDetect(' 你好 ');
})();
```

5.  開一個新的 Terminal 並且執行 node ./intentDetect.js

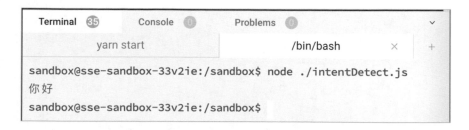

6.  再來要 google 的 API 溝通，需要憑證，而這些都去 GCP 拿

### 9-6-2 拿憑證

我們這裡會從 dialogflow 處，直接開啟 GCP 的連結，去拿憑證，但若你沒有用過 GCP 的話，GCP 會要求你註冊，註冊完成後會給你 $300 美金的 credits 讓你可以抵用。

是的，註冊得交出信用卡，我知道交出信用卡會怕怕，我也希望可以不要，但這目前無解，想要用 Google 的 API 就是得交出信用卡，所以請自行評估。

1.　開啟剛剛的 dialogflow agent

2.　點 ⚙

3.　至 General 頁下方的 GOOGLE PROJECT

4.　點 Project ID 的右邊藍色 ID ，就會開啟 GCP 分頁

5.　左邊選單有一個 API 和服務,點進去

6.　進來 API 和服務頁後

7.　點 啟用 API 和服務

## 8. 會看到很多 Google 服務的 API 列表

## 9. 搜尋 dialogflow API

## 10. 點進去

11. 點管理

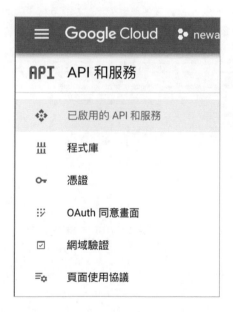

12. 點憑證

13. 點建立憑證

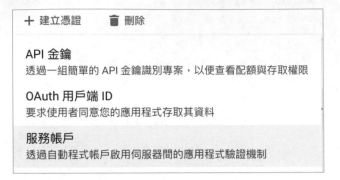

14. 選 服務帳戶

15. 服務帳戶 ID 為必填，這裡我取 demo-dialogflow
16. 點建立並繼續

**❷ 將專案存取權授予這個服務帳戶** (選用)

將「newagent-rjgf」存取權授予這個服務帳戶,讓服務帳戶有權限對專案中的資源完成特定動作。瞭解詳情

| 請選擇角色 ▼ | **條件**<br>新增條件 | 🗑 |

＋ 新增其他角色

繼續

17. 點 選擇角色 ( 這裡指的是要幫這個帳戶開什麼權限 )

請選擇角色 ──────── **條件**                    🗑

☰ 篩選條件 篩選類型

| 快速存取 | 角色 |
| 目前使用 | **Dialogflow API 用戶端** |
| 基本 | Dialogflow 服務代理人 |
| 依產品或服務劃分 | **擁有者** |
| 二進位授權 | |
| 工作流程 | |
| 支援人員 | |

**Dialogflow API 用戶端**
可以針對工作階段、對話資源及其下層項目呼叫所有方法。

管理角色

18. 這裡選 Dialogflow API 用戶端 ( 通常權限開的越限定越好,不要隨便給太高的權限,雖然給高的權限,開發上較方便,但如果權證流出的話,那可就不妙了,此為經驗談 )

＋ 新增其他角色

繼續

19. 點繼續

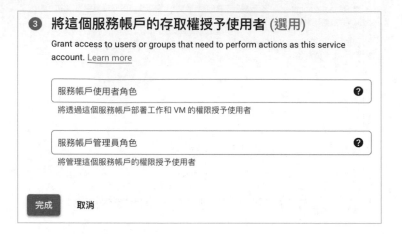

20. 點完成

21. 點進去

22. 中間選項有一個金鑰

23. 點 新增金鑰

24. 點 建立新的金鑰

25. 點 建立

26. 下載了一個 JSON 檔

27. 可以看到 project_id , client_email 及 private_key 都在這裡了

## 9-6-3 使用 dialogflow API

1. 將 project_id , client_email 及 private_key 新增至 Secret keys

目前 intentDetect.js:

```javascript
const dialogflow = require("dialogflow");

const intentDetect = async (text) => {
 console.log(text);
};
module.exports = intentDetect;
/* */
(() => {
 intentDetect(' 你好 ');
})();
```

2. 我們直接安裝官方套件 dialogflow

3. 在 intentDetect 裡引用 dialogflow

```javascript
const dialogflow = require("dialogflow");
```

4. 宣告一個 dialogflow.SessionsClient 實體

```javascript
const sessionClient = new dialogflow.SessionsClient({
  credentials: {
    client_email: process.env.client_email,
    private_key: process.env.private_key.replace(/\\n/g, '\n') // 因為 \
n 是換行符號，但換行符號在儲存成 Secrets key 時，會被轉存成字串，所以必須再換回
換行符號
  }
});
```

5. 建立 sessionPath ，需要一個 Session ID 的參數，Session ID 必須為唯
   一值，所以安裝 uuid 套件

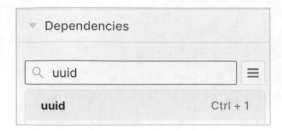

6. 引用 uuid

```
const uuid = require("uuid");
```

7. 建立 sessionPath

```
const sessionPath = sessionClient.sessionPath(
  process.env.project_id,
  uuid.v4()
);
```

8. 建立 request

```
const request = {
  session: sessionPath,
  queryInput: {  // query 發送給 dialogflow agent
    text: {
      text: text,
      languageCode: "zh-TW" // 中文語系
    }
  }
};
```

9. 呼叫 sessionClient.detectIntent 偵測 text 的意圖

```
const responses = await sessionClient.detectIntent(request);
```

10. 可以在 responses 查找到 responses[0].queryResult.intent.displayName 就
是 Intent 的名稱,而 responses[0].queryResult.parameters.fields 就是參
數。

11. 例如：輸入 台北天氣

```
(() => {
 intentDetect(' 台北天氣 ');
})();
```

12. 就會得到意圖為 weather，而參數值為

```
{ city: { stringValue: ' 臺北市 ', kind: 'stringValue' } }
```

13. 我們就可以利用這個來替 LINE Bot 加上大腦了。

# 9-7 dialogflow 練習一

## 9-7-1 改成 npm 套件並發佈

我們呼叫 intentDetect 函式時，期待的 return 值 為：

```
{ intent: 'weather', entites: { city: ' 高雄市 ' } }
```

這樣會比 呼叫 sessionClient.detectIntent 回的傳值，明確也簡單的多。

將這個需求撰寫成套件發布之

建議先自行練習，如果無法完成，再來看答案。

＊ 答案

1. 我們先拿 intent 值，先宣告

```
let intent;
```

2. 拿 intent 值，如果連 intent 值都沒有，就直接報錯

```
try {
  intent = responses[0].queryResult.intent.displayName;
} catch (err) {
  return { error: 'no intent' };
}
```

3. 為了拿 entites 的值，宣告 entites 及擷取 fields

```
let entites = {};
let { fields } = responses[0].queryResult.parameters;
```

4. 如果 fields 的長度為 0，表示沒有值，直接 return intent

```
if (Object.keys(fields).length === 0) {
  return { intent };
}
```

5. 逐一去找出值並加入 entites

```
Object.keys(fields).map((i) => {
  if (fields[i].kind === 'stringValue') { // dialogflow 有依照不同資料別分成不同的回傳值
    entites[i] = fields[i].stringValue; // 直接以 key 值宣告 value
  }else if (fields[i].kind === 'numberValue') {
    entites[i] = fields[i].numberValue;
  }
});
```

6. 完成

```
return { intent, entites };
```

目前的 detectIntent.js:

```
const dialogflow = require( "dialogflow" );
const uuid = require( "uuid" );

const intentDetect = async (text, client_email, private_key, project_id)
=> { // 為了改成 module 所以要改成接參數，故將 client_email, private_key,
project_id 作為參數名稱傳進來
  const sessionClient = new dialogflow.SessionsClient({
    credentials: {
      client_email: client_email || process.env.client_email,  // 有
client_email 就用 client_email，以下皆是
      private_key:
        private_key.replace(/\\n/g, '\n') ||
        process.env.private_key.replace(/\\n/g, '\n')
    }
  });
```

```javascript
  const sessionPath = sessionClient.sessionPath(
    project_id || process.env.project_id,
    uuid.v4()
  );
  const request = {
    session: sessionPath,
    queryInput: {
      text: {
        text: text,
        languageCode: "zh-TW"
      }
    }
  };

  const responses = await sessionClient.detectIntent(request);
  let intent;
  try {
    intent = responses[0].queryResult.intent.displayName;
  } catch (err) {
    return { error: 'no intent' };
  }

  let entites = {};
  let { fields } = responses[0].queryResult.parameters;

  if (Object.keys(fields).length === 0) {
    return { intent };
  }

  Object.keys(fields).map((i) => {
    if (fields[i].kind === 'stringValue') {
      entites[i] = fields[i].stringValue;
    } else if (fields[i].kind === 'numberValue') {
      entites[i] = fields[i].numberValue;
    }
  });

  return { intent, entites };
};
module.exports = intentDetect;
```

7. 修改 index.js

```javascript
module.exports = require("./intentDetect");
```

8. 修改 package.json

```json
{
 "name": "dialogflow-easy-get-value-xxx",
 "version": "1.0.0",
 "license": "MIT",
 "main": "intentDetect.js",
 "dependencies": {
   "dialogflow": "1.2.0",
   "uuid": "8.3.2"
 },
 "devDependencies": {
   "@types/node": "^17.0.21",
   "nodemon": "2.0.19"
 }
}
```

9. npm publish 發佈

# 9-8 dialogflow 練習二

## 9-8-1 結合 LINE Bot 完成天氣小幫手

建議先自行練習，如果無法完成，再來看答案。

＊ 答案

1.　安裝 express , @line/bot-sdk , nodemon

package.json :

```
"dependencies": {
   "@line/bot-sdk": "7.5.0",
   "express": "4.18.1"
 },
 "scripts": {
   "start": "nodemon index.js"
 },
 "devDependencies": {
   "@types/node": "^17.0.21",
   "nodemon": "2.0.19"
 },
```

2.　新增 channelAccessToken , channelSecret 至 Secret Keys

3. 先完成一個 echo bot

目前的 index.js

```javascript
const express = require("express");
const line = require("@line/bot-sdk");
const app = express();
const config = {
 channelAccessToken: process.env.channelAccessToken,
 channelSecret: process.env.channelSecret
};
const client = new line.Client(config);

app.get('/', (req, res) => {
 res.end('hello');
});
app.post("/webhook", line.middleware(config), async (req, res) => {
 let { text } = req.body.events[0].message; // 使用者傳來的訊息
 let msg = {
   type: "text",
   text
  };
 client.replyMessage(req.body.events[0].replyToken, msg);
});
app.listen(8080);
```

我們要作的就是拿到使用者訊息 text 後，去呼叫 dialogflow 去判斷 intent 跟 entites 然後呼叫 氣象局 API 去拿天氣資料，這裡就可以用我們寫的 npm package 去做就可以了

4. 安裝 dialogflow 篇開發的套件 dialogflow-easy-get-value-xxx 並引用

```javascript
const detect = require('dialogflow-easy-get-value-1');
```

5. 呼叫 detect 函式去判斷

```javascript
let meaning = await detect(text);
```

6. 拿值

```javascript
let { intent, entites } = meaning;
```

7. 安裝 開源篇開發的套件 taiwan-weather-xxx 並引用

```
const tw = require('taiwan-weather-1');
```

8. 判斷 intent

```
if (intent === 'weather') { }else{ }
```

9. 若 intent 為 'weather' 就呼叫 tw 函式去拿天氣資料

```
let data = await tw(process.env.weather_key, entites.city);
   console.log(data);
   let { city, degree, weather, feeling, chance } = data;
   text = '${city} 今天的天氣為 ${weather} 感覺 ${feeling} 最高溫為
${degree.max} 度 最低溫為 ${degree.min} 度 降雨機率為 ${chance} % ';
```

10. 若 intent 不是 'weather' 就功能提示

```
text = ' 聽不懂 ${text} 你可以問我 台北天氣 ';
```

目前的 index.js:

```
const tw = require('taiwan-weather-1');
const detect = require('dialogflow-easy-get-value-1');
const express = require('express');
const line = require('@line/bot-sdk');
const CONFIG = {
 channelAccessToken: process.env.channelAccessToken,
 channelSecret: process.env.channelSecret
};
const app = express();
const client = new line.Client(CONFIG);

app.get('/', (req, res) => res.end('hello'));
app.post('/webhook', line.middleware(CONFIG), async (req, res) => {
 let { text } = req.body.events[0].message;
 let meaning = await detect(text);

 let { intent, entites } = meaning;
 console.log(meaning);
```

```
if (intent === 'weather') {
  let data = await tw(process.env.weather_key, entites.city);
  console.log(data);
  let { city, degree, weather, feeling, chance } = data;
  text = '${city} 今天的天氣為 ${weather} 感覺 ${feeling} 最高溫為
${degree.max} 度 最低溫為 ${degree.min} 度 降雨機率為 ${chance} % ';
} else {
  text = ' 聽不懂 ${text} 你可以問我 台北天氣 ';
}
let msg = {
  type: "text",
  text
};

let r = await client.replyMessage(req.body.events[0].replyToken, msg);
res.json(r);
});
app.listen(8080, () => console.log('running'));
```

11. 試一試

## ▍9-9 dialogflow 練習三

### 9-9-1 利用 dialogflow API 將 剪刀石頭布的遊戲邏輯 加進 天 氣小幫手 LINE Bot 裡，又不影響原有的天氣查詢功能

# 10

## 上雲篇
## GCP

## ▋ 10-1 前言

我們到目前為止都是直接使用 codesandbox 作為 LINE Bot 的後端解決方案，但只要 codesandbox 一關掉，我們的 LINE Bot 就跟著 shut down 了，這樣當然是不行的，本篇會以實作的方式，引導讀者將 LINE Bot 上架至 Google Cloud Function 上，這樣我們的 LINE Bot 就有穩定可靠的 backend 後端解決方案了。

## ▋ 10-2 學習目標／演練成果

本章藉由實作 google cloud function 上架前一章的 LINE Bot 來了解何謂後端解決方案。

## ▋ 10-3 部屬到 Google Cloud Function

google cloud function 是 FAAS Functions-as-a-Service

### 10-3-1 啟用 google cloud function

1.  在 GCP 裡搜尋 cloud function

2.  第一個就是 Cloud Functions，點進去

3. 點 建立函式

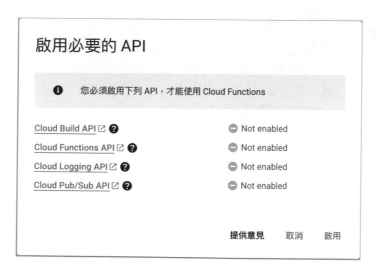

4. 有很多必要的 API 必須先啟動，才能使用 Google Cloud Functions，點 啟用，當然如果你還沒有完成註冊的程序 ( 就是交出信用卡 )，到這裡 應該也是會被擋。

5. 到了建立函式頁

6. 地區改成 asia-east1 也就是台灣彰化的機房,因為台灣彰化離在台灣的我們最近,反應當然會最快。

7. 驗證改成 允許未經驗證的叫用

8. 點 儲存

9. 點 下一步 下一步 取消

10. 下一頁是 程式碼 頁

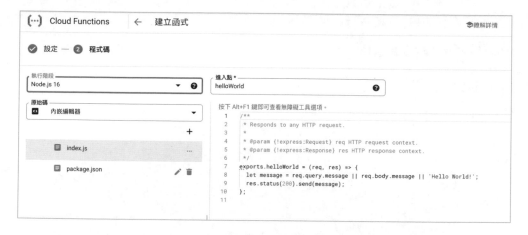

11. 進入點是會呼叫那個函式 預設是 helloWorld

12. 預設的 index.js

```
exports.helloWorld = (req, res) => {
  let message = req.query.message || req.body.message || 'Hello World!';
  res.status(200).send(message);
};
```

helloWorld 會印出 'Hello World!'

13. 預設的 package.json

```
{
  "name": "sample-http",
  "version": "0.0.1"
}
```

沒有安裝任何套件

14. 點部署 部署

| | ● | 環境 | 名稱 ↑ | 區域 | 觸發條件 | 執行階段 | 分配的記憶體 | 已執行的函式 | 最近部署時間 |
|---|---|---|---|---|---|---|---|---|---|
| ☐ | ⟲ | 1st gen | function-1 | asia-east1 | HTTP | Node.js 16 | 256 MB | helloWorld | 2022年7月26日 下午3:10:24 |

Cloud Functions　函式　➕建立函式　⟳重新整理

☰ 篩選　篩選函式

15. 到了下一頁，GCP 正在開資源出來

16. 好了之後 點 function-1

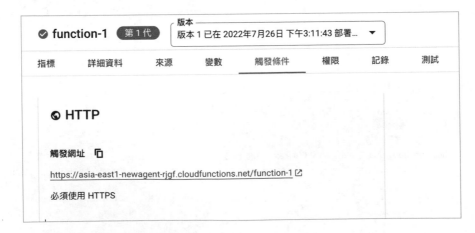

17. 點 觸發條件

18. 點 觸發網址 會另開新頁

19. 就會看到 Hello World! 了

20. 加個 query

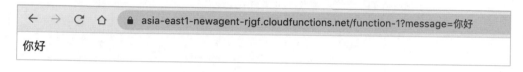

21. 運作正常

## 10-3-2 將 LINE Bot 放上 google cloud function

google cloud function 的觸發方式是，直接呼叫函式，可是我們前面用的是 express 來做 LINE Bot 的 webhook，因此必須要有一些改動來符合 google cloud function 的格式。

我們這裡實作一個簡單的 echo bot

1. 建立函式

2. 函式名稱改成 webhook-echo-bot

3. 地區改成 asia-northeast1 也就是東京的機房，LINE 的機房在東京，這樣會加快回應的速度

4. 驗證改成 允許未經驗證的叫用

5. 點 儲存

6. 點 Runtime, build, connections and security settings

**Runtime, build, connections and security settings** ∧

執行階段　　版本　　連線　　安全性和映像檔存放區

分配的記憶體 *
256 MB

逾時 *
60　　　　　　　　　　　　　　　　　　　　　秒 ❓

**執行階段服務帳戶** ❓

執行階段服務帳戶
App Engine default service account

根據預設，Cloud Functions 會使用系統自動建立的預設 App Engine 服務帳戶。進一步瞭解服務帳戶。

**自動調度資源** ❓

執行個體數量下限
0

執行個體數量上限
3000

**執行階段環境變數** ❓

＋ 新增變數

7. 在 執行階段環境變數 點 新增變數

8. 新增變數 將 Channel secret 跟 Channel access token 的值新增上去

9. 點 下一步 下一步 取消

10. 下一頁是 程式碼 頁

11. 先修改 package.json 加上 @line/bot-sdk 套件

```json
{
  "name": "sample-http",
  "version": "0.0.1",
   "dependencies": {
     "@line/bot-sdk": "7.5.0"
  }
}
```

加上 @line/bot-sdk

12. 進入點改成 webhook

進入點 *
**webhook**                                                    ❓

13. 我們實作一個簡單的 echo bot

index.js :

```javascript
const line = require("@line/bot-sdk");

const config = {
 channelAccessToken: process.env.token,
 channelSecret: process.env.secret
};
const client = new line.Client(config);

exports.webhook = async (req, res) => { //export webhook 出去為 進入點
 let { text } = req.body.events[0].message;
 let msg = {
   type: "text",
   text
 };
 let r = await client.replyMessage(req.body.events[0].replyToken, msg);
 res.json(r);
};
```

原本是用 express，但為了符合 google cloud function 的規則，故改成

```
exports.webhook = async (req, res) =>{...}
```

14. 點 部署 部署

15. 好了之後 點 webhook-echo-bot

16. 點 觸發條件

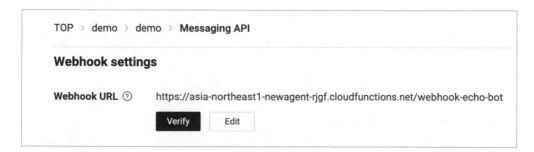

17. 將 觸發網址 貼到 LINE Bot 的 webhook url 設定

18. 試一試

## 10-3-3 上線天氣小幫手

以上一節為基礎 修改後 上線至 gcp

1. 勾起來後，點複製

2. 改名稱

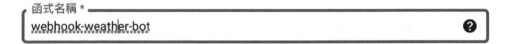

3. 選 允許未經驗證的叫用

4. 複製的好處就是剛剛的環境變數不用再填寫

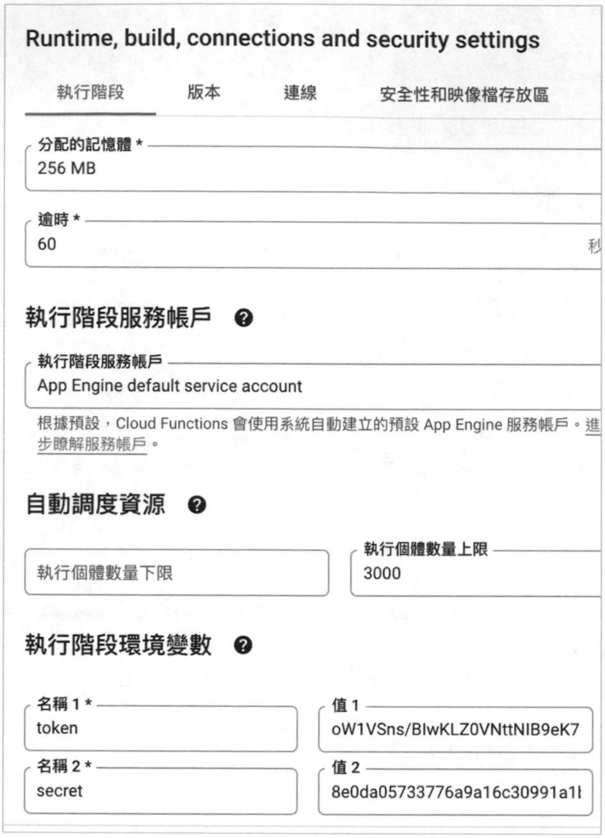


5. 再加上 dialogflow API 以及 氣象局 open data 需要的 key 值

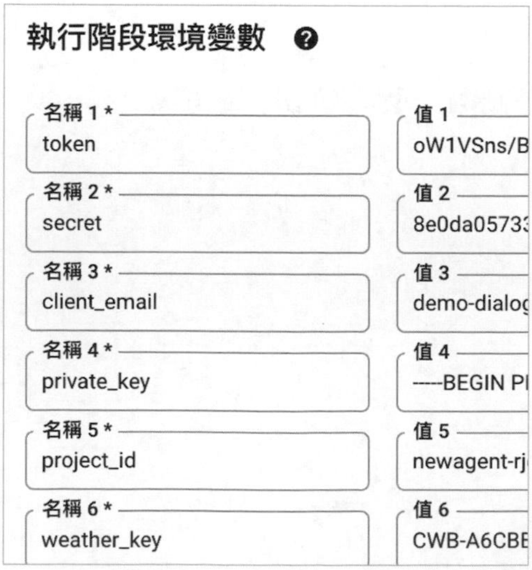

6.   點儲存

7.   按下一步

8.   修 改 package.json 加 上 dialogflow-easy-get-value-xxx, taiwan-weather-
     xxx 套件

```
"dependencies": {
  "@line/bot-sdk": "7.5.0",
  "dialogflow-easy-get-value-1": "1.0.3",
  "taiwan-weather-1": "1.0.5"
},
```

9.   修改 index.js 的進入點 webhook 的程式碼

```
exports.webhook = async (req, res) => { //export webhook 出去為 進入點
 let { text } = req.body.events[0].message;
  console.log(text);
 let meaning = await detect(text);

 let { intent, entites } = meaning;
  console.log(meaning);
 if (intent === 'weather') {
   let data = await tw(process.env.weather_key, entites.city);
   console.log(data);
   let { city, degree, weather, feeling, chance } = data;
   text = '${city} 今天的天氣為 ${weather} 感覺 ${feeling} 最高溫為
${degree.max} 度 最低溫為 ${degree.min} 度 降雨機率為 ${chance} % ';
 } else {
   text = ' 聽不懂 ${text} 你可以問我 台北天氣 ';
 }
 let msg = {
   type: "text",
   text
 };

 let r = await client.replyMessage(req.body.events[0].replyToken, msg);
  res.json(r);
};
```

10.  點部署

## 11. 試一試

目前的 index.js:

```javascript
const tw = require('taiwan-weather-1');
const detect = require('dialogflow-easy-get-value-1');
const line = require("@line/bot-sdk");

const config = {
channelAccessToken: process.env.token,
channelSecret: process.env.secret
};
const client = new line.Client(config);

exports.webhook = async (req, res) => { //export webhook 出去為 進入點
let { text } = req.body.events[0].message;
 console.log(text);
 let meaning = await detect(text);

 let { intent, entites } = meaning;
 console.log(meaning);
 if (intent === 'weather') {
   let data = await tw(process.env.weather_key, entites.city);
   console.log(data);
   let { city, degree, weather, feeling, chance } = data;
   text = '${city} 今天的天氣為 ${weather} 感覺 ${feeling} 最高溫為
${degree.max} 度 最低溫為 ${degree.min} 度 降雨機率為 ${chance} % ';
 } else {
   text = ' 聽不懂 ${text} 你可以問我 台北天氣 ';
 }
 let msg = {
   type: "text",
   text
 };

 let r = await client.replyMessage(req.body.events[0].replyToken, msg);
 res.json(r);
};
```

## ▌ 10-4 附帶一提：後端解決方案歷史

本章是採用目前最流行的 serverless 的概念，所衍伸出的 google cloud function 的後端解決方案來作為實作。

但後端解決方案走到 serverless 概念的今天，並不是一蹴可幾的，是經過好幾年不斷演進的。

這裡將就後端解決方案的歷史，做一個介紹。

### 10-4-1 本機機房

最早期當筆者接觸網站開發時，那時瀏覽器問世沒多久，是 web 1.0 的時代。

那個時候所謂架網站，就是用一台電腦，安裝伺服軟體，當時流行的是 Apache ，運行好 Apache 之後，在預設的資料裡置入網頁檔案，並確保網路通暢，讓外面的電腦可以藉由網路，連到你的電腦之後，再藉由瀏覽器去打開伺服軟體裡的的網頁檔案，這就是所謂架設網站了。

但這個方式有幾個問題

1. 電腦除了網路伺服之外，因為是個人電腦，還會去運行其他軟體
2. 可能一不注意，就踢到線，造成電腦 shut down ，網站也就跟著 shut down

### 10-4-2 代管機房

當然上述問題的解決，可以買一台電腦當專用網站伺服，並置於固定機房中。

但是有這個需求的不會只有一間公司，因此就有代管機房的服務出現。

你只要將你的電腦放到他們的機房就可以了，而他們保證電腦供應正常以及網路的暢通。

### 10-4-3 PAAS (platform as a service) 平台即服務

可是電腦搬來搬去也不是個辦法，其實使用者在意的只有服務是否正常，並不想去維護硬體和作業系統等細節。

因此就有平台提供程式的方式，讓使用者將開發好的網路服務上傳並運行至該平台，專注在自己的業務邏輯即可。

而 aws elastic beanstalk 和 google app engine 都是當時有名的 PAAS 服務。

### 10-4-4 Microservices 微服務

隨著現代人對於網路依賴，我們發現有的時候，當網站某個功能運行失效，可能導致整個網站的 shut down，但網站的其他功能不應該跟著 shut down，因此有人提出，可不可以將網站拆分成許多服務來避免一個功能失效時，導致所有功能跟著失效。

而此時 docker 容器化的問世，又更進一步加速這個目標的實現。

主要實作做法就是將網站上的各個服務，拆成各個對應的容器，例如物流、金流等。

這樣做的好處，除了避免一個功能失效造成整個網站 shut down 之外；還可以分開各個功能的迭代更新，不需要每次更新，都要整個網站全部更新；還可以監控那些容器佔據較多資源，那些容器較無人使用，可以讓資源做更有效的分配及整理。

而 Kubernetes / k8s 是目前的主流管控軟體。

### 10-4-5 FAAS Function-as-a-Service 功能即服務

又被稱為無伺服器運算 ，也就是目前最新的概念。

主要實作做法就是將服務拆成各個函式上傳至服務後，以 web api 的方式去呼叫，好處是連維運都省了。

不過要將所有功能都拆分成函式,相當考驗工程師的功力。

目前有提供這類服務的有 google cloud function、aws lambda、azure function。

## ▌ 10-5 練習

1. 試將攝式華式換算上架 google cloud function

2. 試將剪刀石頭布遊戲上架 google cloud function

# 11

## 總結篇
## JS 的無限可能

Anything that can be Written in JavaScript, will Eventually be Written in JavaScript - by John M. Wrigh

任何可以用 JavaScript 實現的，都將被 JavaScript 來實現。

終於看了這裡，對於 JS 有點概念了，那下一步呢？

下一步，對於初學者來說，當然是寫寫 Side Project 來練習對於 JS 的熟悉度。

動手使用 JS 程式改善改進你現在的生活模式，往往是一個使自己程式能力進步的好方法。

可喜的是 Javascript 在 github 上已經蟬聯了 8 年第一名的程式語言，在各式的應用場景，都可以找到 JS 的 framework ，也就是說不管你想開發什麼應用，通通找得到資源。

# ▋ 11-1　後端解決方案

node.js 誕生的初心，本就是因為作者覺得為什麼為了做網站，必須要學兩套語言，前端是 JS ，後端是 PHP ，所以他就創造了 node.js 想取代後端的 PHP ，後來更衍生出 NPM ，狀大了整個 node.js 社群。

然後，就你所看到的 node.js 的各式應用情景的 framework 越來越多了。

有那些後端項目的 Side Project ，你可以動手作呢？

## 11-1-1　Website Backend

這個不用說，就是 node.js 作者的初心，使用 express , koa 等套件，就能作為 website 的 backend 了，但是你還得學會前端開發的 HTML , CSS 等知識。

以下是使用 Express 的基本範例。

```javascript
const express = require('express')
const app = express()
const port = 3000

app.get('/', (req, res) => {
  res.send('Hello World!')
})

app.listen(port, () => {
  console.log('Example app listening on port ${port}')
})
```

## 11-1-2 Chatbot Backend

聊天機器人（ChatBot）是經由對話或文字進行交談的電腦程式來模擬人類對話。 聊天機器人可以作為，客戶服務或資訊獲取。聊天機器人可以搭載自然語言處理系統，例如 Dialogflow 去協助使用者的意圖。

這也是本書另一個著重的部份，就是 Chatbot ，寫 Chatbot 的好處就是不需要去懂得太多 knowledge base 例如 HTML , CSS 時，就可以撰寫 Chatbot 了。

以下是使用 Express 的架設 LINE Bot 伺服器的基本範例。

```javascript
const express = require('express');
const line = require('@line/bot-sdk');

const config = {
  channelAccessToken: 'YOUR_CHANNEL_ACCESS_TOKEN',
  channelSecret: 'YOUR_CHANNEL_SECRET'
};

const app = express();
app.post('/webhook', line.middleware(config), (req, res) => {
  Promise
    .all(req.body.events.map(handleEvent))
    .then((result) => res.json(result));
});
```

```
const client = new line.Client(config);
function handleEvent(event) {
  if (event.type !== 'message' || event.message.type !== 'text') {
    return Promise.resolve(null);
  }

  return client.replyMessage(event.replyToken, {
    type: 'text',
    text: event.message.text
  });
}

app.listen(3000);
```

## 11-1-3　Web API

API 應用程式介面（application programming interface），是一種介面，定義軟體中介之間的互動，以及可以進行的呼叫（call）或請求（request）等。

從 API 去著手是一個很好的選項，像是本書中，將氣象局天氣的資訊，依照自己的需求更改為簡單的格式。

你也可以試試依照自己的需求，動手去撰寫自己需要的 API。

## 11-1-4　GraphQL

GraphQL 是資料庫查詢的衍生應用，有別以前的 SQL 語法，使用 GraphQL 可以自定義自己的查詢，可以一次的將所有要查詢或是變更的內容定義在同一個查詢裡，藉此可以減少查詢的次數，是目前流行的資料查詢操作語言。

以下是 GraphQL 的基本語法：

```
query list($page: Number!) {
  postList(page: $page) {
    id
    title
    content
    createdAt
  }
}
```

## ▋ 11-2　前端網頁

在最早的網站架設三本柱裡，就是 apache + php + mysql，前端在當時，不過就是 HTML 檔案，放在 apache 裡的一個資料夾，檔案名稱 index.html 就是網站開啟的首頁，要開啟其他的網頁，則要做好超連結，這樣就是網站了。

但很快這個結構就不夠用了，為了各種需求，例如：多種螢幕 size，出了個響應式網頁等。

漸漸的人們不再撰寫純 HTML 檔案，也無法只撰寫純 HTML，來滿足目前網站開發的需求，因而衍伸出了三種主要的前端網頁開發框架，以及其衍伸套件。

三種套件各有支持者，你可以選擇自己喜歡的開發框架來撰寫網頁。

### 11-2-1　React

React（也稱為 React.js 或 ReactJS）是一個免費的開放原始碼前端 JavaScript 工具庫，是基於 UI 組件構建使用者介面，主要是由 Meta 前身是 facebook 所維護以下是使用 JSX 和 JavaScript 在 HTML 中使用 React 的基本範例。

```
import React from "react";
const Greeting = () => {
 return (
   <div className="hello_world">
     <h1> Hello, world! </h1>
   </div>
 );
};
export default Greeting;
```

## 11-2-2 Vue

Vue.js，簡稱 Vue，是一個用於建立使用者介面的開源 MVVM 前端 JavaScript 框架，也是一個建立單頁應用的 Web 應用框架。它由尤雨溪建立，由他和其他活躍的核心團隊成員維護。

以下是使用 Vue 的基本範例。

```
import { createApp } from 'vue'

createApp({
  data() {
    return {
      count: 0
    }
  }
}).mount('#app')
```

```
<div id="app">
  <button @click="count++">
    Count is: {{ count }}
  </button>
</div>
```

## 11-2-3 angular

AngularJS 是一款由 Google 維護的開源 JavaScript 函式庫。用來協助單一頁面應用程式運行,目前已停止維護。

# 11-3 區塊鏈

## 11-3-1 web3.js

Web3(也被稱為 Web 3.0,又寫為 web3[1])是關於全球資訊網發展的一個概念,主要與基於區塊鏈的去中心化、加密貨幣以及非同質化代幣有關。

以下是使用 web3.js 的基本範例。

```javascript
// In Node.js
const Web3 = require('web3');
const web3 = new Web3('ws://localhost:8546');
console.log(web3);
// Output
{
    eth: ... ,
    shh: ... ,
    utils: ...,
    ...
}
```

```javascript
web3.setProvider('ws://localhost:8546');
// or
web3.setProvider(new Web3.providers.WebsocketProvider('ws://
localhost:8546'));
```

```javascript
web3.eth.getAccounts().then(console.log);
```

### 11-3-2 NFT

非同質化代幣（英語：Non-Fungible Token，簡稱：NFT），是一種被稱為區塊鏈數位帳本上的資料單位，每個代幣可以代表一個獨特的數位資料，作為虛擬商品所有權的電子認證或憑證。由於其不能互換的特性，非同質化代幣可以代表數位資產，如畫作、藝術品、聲音、影片、遊戲中的專案或其他形式的創意作品。雖然作品本身是可以無限複製的，但這些代表它們的代幣在其底層區塊鏈上能被完整追蹤，故能為買家提供所有權證明。諸如以太幣、比特幣等加密貨幣都有自己的代幣標準以定義對 NFT 的使用。

可以在 web3 裡，找到發行 NFT 的教程。

## ▌11-4 人工智慧

人工智慧（英語：artificial intelligence，縮寫為 AI）亦稱智械、機器智慧，指由人製造出來的機器所表現出來的智慧。通常人工智慧是指透過普通電腦程式來呈現人類智慧的技術。該詞也指出研究這樣的智慧系統是否能夠實現，以及如何實現。同時，透過醫學、神經科學、機器人學及統計學等的進步

人工智慧的四個主要組成部分是：

- 專家系統：作為專家處理正在審查的情況，並產生預期的績效。
- 啟發式問題解決：包括評估小範圍的解決方案，並可能涉及一些猜測，以找到接近最佳的解決方案。
- 自然語言處理：在自然語言中實現人機之間的交流。
- 計算機視覺：自動生成識別形狀和功能的能力

### 11-4-1 tensorflow.js

TensorFlow 是一個開源軟體庫，用於各種感知和語言理解任務的機器學習。目前被 50 個團隊用於研究和生產許多 Google 商業產品，如語音辨識、Gmail、Google 相簿和搜尋。

以下是使用 tensorflow.js 的基本範例。

```
import * as tf from '@tensorflow/tfjs';
const model = await tf.loadLayersModel('https://foo.bar/tfjs_artifacts/
model.json');
```

```
const example = tf.fromPixels(webcamElement);  // for example
const prediction = model.predict(example);
```

### 11-4-2 人工智慧相關服務 API

例如本書介紹用於自然語言處理的 Google dialogflow 有 npm 的 package 可以作開發外，除了 Google 之外，還有其他的廠商、林林總總的不勝枚舉。

## 11-5 IOT

Internet of Things 物聯網，是硬體裝置相互聯結的系統，透過網路傳輸收集到的數據之用。

常見的有偵測溫度、溼度、物體偵測、開關等。

### 11-5-1 Node-RED

Node-RED 是 IBM 以 Node.js 為基礎所開發出來的 IOT 系統

以下是使用 Node-RED 的基本範例。

```
if(msg.payload.result == true){
    msg.payload.note = "yes"
}
return msg;
```

## 11-6 APP 行動應用程式

手機 app 的開發，也已過了十多個年頭了，從早期 android 只能使用 java 開發，mac 只能使用 object-c 開發，到現在很多語言都支援 APP 開發，可謂是百花齊放。

### 11-6-1 React Native

React Native 是 FB 基於 React.js 所開發的應用程式架構，讓開發者可以利用 JavaScript 和 React.js 的開發模式開發出 APP。

以下是使用 React Native 的基本範例。

```jsx
import React from 'react';
import {Text, View} from 'react-native';
import {Header} from './Header';
import {heading} from './Typography';

const WelcomeScreen = () =>
  <View>
    <Header title="Welcome to React Native"/>
    <Text style={heading}>Step One</Text>
    <Text>
      Edit App.js to change this screen and turn it
      into your app.
    </Text>
    <Text style={heading}>See Your Changes</Text>
    <Text>
      Press Cmd + R inside the simulator to reload
      your app's code.
    </Text>
    <Text style={heading}>Debug</Text>
    <Text>
      Press Cmd + M or Shake your device to open the
      React Native Debug Menu.
    </Text>
    <Text style={heading}>Learn</Text>
    <Text>
      Read the docs to discover what to do next:
    </Text>
  </View>
```

## ▋ 11-7 桌面應用軟體

### 11-7-1 Electron

Electron 透過使用 Node.js 和 Chromium engine 完成跨平台的桌面 GUI 應用程式的開發。

以下是使用 Electron 的基本範例。

main.js：

```javascript
const { app, BrowserWindow } = require('electron')
const path = require('path')

function createWindow () {
  const win = new BrowserWindow({
    width: 800,
    height: 600,
    webPreferences: {
      preload: path.join(__dirname, 'preload.js')
    }
  })

  win.loadFile('index.html')
}

app.whenReady().then(() => {
  createWindow()

  app.on('activate', () => {
    if (BrowserWindow.getAllWindows().length === 0) {
      createWindow()
    }
  })
})

app.on('window-all-closed', () => {
  if (process.platform !== 'darwin') {
    app.quit()
  }
})
```

preload.js：

```javascript
window.addEventListener('DOMContentLoaded', () => {
  const replaceText = (selector, text) => {
    const element = document.getElementById(selector)
    if (element) element.innerText = text
  }

  for (const type of ['chrome', 'node', 'electron']) {
    replaceText('${type}-version', process.versions[type])
  }
})
```

index.html

```html
<!DOCTYPE html>
<html>
<head>
    <meta charset="UTF-8">
    <title>Hello World!</title>
    <meta http-equiv="Content-Security-Policy" content="script-src 'self'
'unsafe-inline';" />
</head>
<body>
    <h1>Hello World!</h1>
    <p>
        We are using Node.js <span id="node-version"></span>,
        Chromium <span id="chrome-version"></span>,
        and Electron <span id="electron-version"></span>.
    </p>
</body>
</html>
```

## ▋ 11-8 自動化測試及爬蟲

自動化測試

隨著軟體系統規模的日益擴大及拓展，對軟體系統的測試也變得更加困難和複雜；故自動測試可以減少人工測試，可能發生的測試疏漏以及重複工作所導致的人為差錯。

爬蟲

那因為自動化測試的環境跟所謂爬蟲需要騙過系統的偵測的需要是一致的，所以使用的 package 也是相同的。

## 11-8-1 selenium-webdriver

Selenium 是一個 Browser 的自動化函式庫。

以下是使用 selenium-webdriver 的基本範例。

```
const {Builder, Browser, By, Key, until} = require('selenium-webdriver');

(async function example() {
  let driver = await new Builder().forBrowser(Browser.FIREFOX).build();
  try {
    await driver.get('http://www.google.com/ncr');
    await driver.findElement(By.name('q')).sendKeys('webdriver', Key.
RETURN);
    await driver.wait(until.titleIs('webdriver - Google Search'), 1000);
  } finally {
    await driver.quit();
  }
})();
```

## 11-8-2 puppeteer

Puppeteer 是一個 Node 函式庫，透過 API 控制 Chrome

以下是使用 Puppeteer 的基本範例。

```
const puppeteer = require('puppeteer');

(async () => {
  const browser = await puppeteer.launch();
  const page = await browser.newPage();
  await page.goto('https://example.com');
  await page.screenshot({path: 'example.png'});

  await browser.close();
})();
```

## 11-9 其他

在 npm 和 github 裡可以找到很多有趣的 package 專注在各式不同的領域，幾乎你想的到的，想不到的都有前人開發及分站過了，建議在寫任何 side project 之前，都可以先在 github 上找找，會有很多意想不到的收獲。

## 11-10 以 JS 為基礎發展的語法等

### 11-10-1 Google Apps Script

Apps Script 是基於 JS 語法，所開發的語法，主要目的是讓開發者創建與 G Suite 集成的業務應用程序。

以下是使用 Apps Script 的基本範例。

```
function myFunction() {
  Browser.msgBox("Hello World!");
}
```

### 11-10-2 Google Ads scripts

Google Ads scripts 提供你藉由簡單的 JS 語法和 Google Ads 廣告帳戶溝通互動。

以下是使用 Google Ads scripts 的基本範例。

```
function main() {
    let keywords = AdsApp.keywords()
        .orderBy("metrics.impressions DESC")
        .forDateRange("YESTERDAY")
        .withLimit(10)
        .get();

    console.log("The 10 keywords with the most impressions yesterday:");
    for (const keyword of keywords) {
        console.log('${keyword.getText()}: ${keyword.
getStatsFor("YESTERDAY")
                                            .getImpressions()}');
    }
}
```

## 11-10-3 TypeScript

TypeScript 可以視為 JS 的強型別加強版,主要目的也是為了補足弱型別言,沒有型別檢查,在開發上的不便;但每次運行都必須透過 TypeScript 編譯器,轉譯成 JS 才能被執行。

以下是使用 TypeScript 的基本範例。

```
let message: string = 'Hello, World!';
console.log(message);
```

## 11-10-4 JSON

JSON:JavaScript Object Notation 是一種輕量級資料交換格式。其內容由屬性和值所組成,因此也有易於閱讀和處理的優勢。目前已被廣泛運用於各種資料的 API 中,如本書所介紹的中央氣象局開放資料平臺,就有提供 JSON 的格式。

以下是使用 中央氣象局開放資料平臺 API 的 JSON 的基本範例。

```json
{
  "success": "true",
  "result": {
    "resource_id": "F-C0032-001",
    "fields": [
      {
        "id": "datasetDescription",
        "type": "String"
      },
      {
        "id": "locationName",
        "type": "String"
      },
      {
        "id": "parameterName",
        "type": "String"
      },
      {
        "id": "parameterValue",
        "type": "String"
      },
      {
        "id": "parameterUnit",
        "type": "String"
      },
      {
        "id": "startTime",
        "type": "Timestamp"
      },
      {
        "id": "endTime",
        "type": "Timestamp"
      }
    ]
  },
```

```
"records": {
  "datasetDescription": " 三十六小時天氣預報 ",
  "location": [
    {
      "locationName": " 宜蘭縣 ",
      "weatherElement": [
        {
          "elementName": "PoP",
          "time": [
            {
              "startTime": "2022-08-10 12:00:00",
              "endTime": "2022-08-10 18:00:00",
              "parameter": {
                "parameterName": "10",
                "parameterUnit": " 百分比 "
              }
            },
            {
              "startTime": "2022-08-10 18:00:00",
              "endTime": "2022-08-11 06:00:00",
              "parameter": {
                "parameterName": "20",
                "parameterUnit": " 百分比 "
              }
            },
            {
              "startTime": "2022-08-11 06:00:00",
              "endTime": "2022-08-11 18:00:00",
              "parameter": {
                "parameterName": "20",
                "parameterUnit": " 百分比 "
              }
            }
          ]
        }
      ]
    }
  ]
}
}
```

## ▌ 工具

工欲善其事,必先利其器。

軟體開發時至有很多免費且開源工具是我們開發不管是哪一種語言,或是哪一種應用,或是哪一種專業領域,都不得不會的工具軟體。

### IDE

IDE 整合開發環境 : Integrated Development Environment,也稱為 Integration Design Environment、Integration Debugging Environment

協助程式開發人員開發軟體的應用軟體,在開發工具內部就可以輔助編寫原始碼文字、並編譯打包成為可用的程式。

本書採用的是 codesandbox ,但到後來如果程式專案變得較複雜時,就得考慮使用其他編輯器了,以前筆者剛接觸時,主流 IDE 都是付費軟體才堪用,但目前免費的就很強大了。

### VS Code

Visual Studio Code: VS Code 是由微軟開發且跨平台的免費原始碼編輯器。

不同於過往的免費 IDE,其將擴充元件程式開放給開發者自行撰寫,並可發佈於擴充元件商店上。

使得其雖然是免費,但功能相較付費 IDE ,一點也不遜色。

### 版本控制

版本控制 Version control 是一種軟體工程技巧,藉此能在軟體開發的過程中,確保由不同人所編輯的同一程式檔案都得到同步。

筆者以前剛上班時,還是以 Subversion 為主 ,不過現在大概就是 git 的天下了。

## ＊ Git

Git 是一個分散式版本控制軟體。

以下是使用 Git 的基本範例。

```
git init
git add
git commit
```

## DevOps 多人協作

DevOps 是 Development 和 Operations 的組合詞，指的是希望 Dev：軟體開發人員和 Ops：IT 運維技術人員之間的溝通合作，但不僅限於這兩者，只要有助於團隊持續進步的，都是 DevOps 所想要實現的目的。

而為了實現這個目的所開發出的各種應用軟體，都是 DevOps 的範疇。

## ＊ Github

github 一開始就真的只是 source code 代管服務，但後來實現了 PR 等功能，逐漸成為多人協作所不可或缺的一份子。

# ▌ Javascript 學習的起點

當然就是本書囉，這有什麼好懷疑的。

## Javascript 學習的終點

學習沒有終點。 XD。最後，僅以本書，獻給一直在學習路上的你跟我。

# 附錄：

參考資料

本書內容參考資料甚多，如有遺漏，敬請見諒。

## ▋ 參考書目

動手學 GitHub ！現代人不能不知道的協同合作平台

平台經濟模式

## ▋ 參考網站

https://zh.wikipedia.org/wiki/

https://www.w3schools.com/js/

https://developer.mozilla.org/en-US/docs/Web/JavaScript

https://kknews.cc/zh-tw/education/

https://expressjs.com/

https://line.github.io/line-bot-sdk-nodejs/